Inventio est excogitation rerum verarum aut veri similium quae causam probabilem reddant[1]

Marcus Tullius Cicero
(~87 BC)

Invention, in Rhetoric, signifies the finding out, and the Choice of Arguments which the Orator is to use for the proving of his Point, or the moving of his Hearers Passions

Ephraim Chambers (1728, p. 402)

Invention:

1. The action of coming upon or finding; the action of finding out; discovery (whether accidental, or the result of search and effort). Rhetoric: the finding out or selection of topics to be treated, or arguments to be used
2. The action of devising, contriving, or making up; contrivance, fabrication
3. The original contrivance or production of a new method or means of doing something, of an art, kind of instrument, etc. previously unknown; origination, introduction. In art and literary composition: the devising of a subject, idea, or method of treatment, by exercise of the intellect or imagination
4. The faculty of inventing or devising; power of mental creation or construction; inventiveness
5. The manner in which a thing is devised or constructed; invented style, fashion, design.

Oxford English Dictionary (2021)

---

1 *'Invention is the discovery of valid or seemingly valid arguments to render one's case plausible'*. From the English edition translated by H.M. Hubbell and published by Harvard University Press/William Heinemann Ltd, Cambridge/London in 1949.

The Invention of Disaster

This theoretical contribution argues that the domination of Western knowledge in disaster scholarship has allowed normative policies and practices of disaster risk reduction to be imposed all over the world. It takes a postcolonial approach to unpack why scholars claim that disasters are social constructs while offering little but theories, concepts and methods supposed to be universal in understanding the unique and diverse experiences of millions of people across very different cultures. It further challenges forms of governments inherited from the Enlightenment that have been rolled out as standard and ultimate solutions to reduce the risk of disaster. Ultimately, the book encourages the emergence of a more diverse set of world views/senses and ways of knowing for both studying disasters and informing policy and practice of disaster risk reduction. Such pluralism is essential to better reflect local realities of what disasters actually are around the world.

This book is an essential read for scholars and postgraduate students interested in disaster studies as well as policy-makers and practitioners of disaster risk reduction.

JC Gaillard is Professor of Geography at Waipapa Taumata Rau (The University of Auckland), Aotearoa (New Zealand).

Routledge Studies in Hazards, Disaster Risk and Climate Change

Series Editor: Ilan Kelman, *Professor of Disasters and Health at the Institute for Risk and Disaster Reduction and the Institute for Global Health, University College London*

This series provides a forum for original and vibrant research. It offers contributions from each of these communities as well as innovative titles that examine the links between hazards, disasters and climate change to bring these schools of thought closer together. This series promotes inter-disciplinary scholarly work that is empirically and theoretically informed, and titles reflect the wealth of research being undertaken in these diverse and exciting fields.

Disaster Deaths
Trends, Causes and Determinants
Bimal Kanti Paul

Disasters and Life in Anticipation of Slow Calamity
Perspectives from the Colombian Andes
Reidar Staupe-Delgado

The Invention of Disaster
Power and Knowledge in Discourses on Hazard and Vulnerability
JC Gaillard

Earthquakes and Volcanic Activity on Islands
History and Contemporary Perspectives from the Azores
David K. Chester, Angus Duncan, Rui Coutinho and Nicolau Wallenstein

For more information about this series, please visit https://www.routledge.com/Routledge-Studies-in-Hazards-Disaster-Risk-and-Climate-Change/book-series/HDC.

The Invention of Disaster

Power and Knowledge in Discourses
on Hazard and Vulnerability

JC Gaillard
Foreword by Noel Castree

Routledge
Taylor & Francis Group

LONDON AND NEW YORK

Cover image: 'Swirls n Spins' by Cyrene Baun-Gaillard, 2014

First published 2022
by Routledge
2 Park Square, Milton Park, Abingdon, Oxon OX14 4RN, UK

and by Routledge
605 Third Avenue, New York, NY 10158, USA

Routledge is an imprint of the Taylor & Francis Group, an informa business

© 2022 JC Gaillard

British Library Cataloguing-in-Publication Data
A catalogue record for this book is available from the British Library

Library of Congress Cataloging-in-Publication Data
A catalog record has been requested for this book

ISBN: 978-1-138-80562-0 (hbk)
ISBN: 978-1-032-16272-0 (pbk)
ISBN: 978-1-315-75216-7 (ebk)

DOI: 10.4324/9781315752167

Typeset in Baskerville
by SPi Technologies India Pvt Ltd (Straive)

Contents

Illustrations

Figures

Tables

Foreword
The power of thinking in disaster studies and disaster risk reduction

JC Gaillard has written a book to be reckoned with. Almost nothing seems as viscerally real as a disaster: tsunamis, earthquakes, volcanoes and the like continue to blight the lives of millions of people worldwide. For instance, at the time of writing (January 2021), floods have killed 20 people and badly affected many localities in the Indonesian province of South Kalimantan. Far worse is to come this year. In the early twenty-first century, there's a huge, multi-scalar apparatus dedicated to understanding disasters and ameliorating their impacts. Hundreds of thousands of researchers, aid workers, government officials, charity workers and health professionals are involved in this deadly serious business. However, around 40 years ago, a few people began to question the realist credentials of expert knowledge in the field of disaster analysis and management. Among them was Ken Hewitt, whose edited book *Interpretations of Calamity* (1983) was formative for many members of my generation. Even the most rigorous researchers and most experienced practitioners were, so Hewitt and his contributors claimed, unable to see disasters through objective eyes. Slowly but surely, what followed was a sustained critique of the 'naturalness' of disasters and a new attention to the socio-economic context that allowed some natural hazards to become disastrous for certain people (but by no means all). The focus on uneven patterns of human vulnerability (and later on resilience) saw a broadening of disaster risk reduction and response beyond technical interventions. There was, we learned, more than one way to see the realities of disasters.

But critique and reflexivity are never guaranteed to become part of the DNA of any field of knowledge and practice. There are many good – and not so good – reasons why thinking congeals and ways of acting get habitualised. In this book, JC Gaillard identifies and questions the many components of expert 'common sense' now prevailing in the multi-disciplinary, multi-agency, multi-national field of disaster studies and management. Like Hewitt et al. four decades ago, he is issuing a calling-out and a call-to-arms. As a member of the field that he anatomises, Gaillard argues that it conflates representation with reality. More pointedly, he makes two plenary claims. The first is that even the critical-Left theory that reverberated through the disaster field from the mid-1980s onwards is actually part of an episteme it ostensibly opposes. The second is that this Western and broadly 'rationalist' episteme has steadily been globalised via United Nations agencies and other means. Gaillard thus throws down the gauntlet to disaster specialists of many different stripes across the world: 'don't simply look in the mirror but *behind* the mirror', he seems to say. The axiomatic needs to be rendered problematic.

Readers will find here a spirited marshalling of postcolonial and more recent decolonial thinking. No one else has synthesised this thinking in order to see the disaster studies and management field anew. As the author notes, 'It has ... never really trickled through the otherwise porous borders of disaster studies and there are very few existing attempts to consider ... disaster, disaster studies and disaster risk reduction from a postcolonial perspective'. Gaillard's is thus a unique intervention and destined to elicit a reaction. Some readers will find his sweeping analysis too confident and encompassing. Others might worry that he's a 'constructionist' who foolishly denies the realities of disasters (for the record, he absolutely does not). Still others will find the book to be inspiring and provocative in the best sense. Whatever the reaction, one hopes that a variety of readers will engage with the book, even those who are somewhat impatient with 'ivory tower' critical theory.

Gaillard combines the sensibilities of a field-worker possessed of long research experience in the disaster arena with those of an intellectual able to stand back and see the contours of a larger landscape of thinking. In these pages, we find discussions of Derridean philosophy, Foucauldian genealogy and postcolonial theory brought to bear on the realities of hazardous lives in the Philippines, Indonesia and Samoa. Gaillard's analysis is impressively wide-ranging. Even in the most progressive nooks and crannies of the disaster field of knowledge and practice, he shows, the forces of colonial reasoning are at work. In short, the book is not an exercise in 'external critique' written by an outsider lacking any sense of the realities of the disaster field. While postcolonial writings and poststructuralist discourse can be painfully esoteric (even obscurantist), Gaillard seeks to show their practical relevance to the field in question.

It would be plain wrong – let alone irresponsibly hyperbolic – to say that the 'Western episteme' has been a disaster for the millions of people who have been affected by its cognitive and normative precepts. Gaillard has no truck with such silliness. But his powerful analysis suggests the urgent need to let other ways of knowing and acting through the door. If structures of thought have a power irreducible to the realities they claim to represent, then deconstructing and recomposing them can empower those currently marginalised in our far-from-perfect world. A revolution in the disasters field may not be in the offing. But this book may help speed the field's evolution towards a more inclusive, tolerant and just *modus operandi.* Disasters kill, maim and damage. But so do the epistemic and practical dimensions of the disaster risk reduction apparatus when shielded from deep critical scrutiny.

Noel Castree
University of Technology Sydney, Australia
University of Manchester, United Kingdom

Preface
On worlds and words

Critical disaster studies is ragged with a baffling paradox. It claims that disasters are social constructs. Yet it offers little but theories, concepts and methods supposed to be universal in their relevance and application, to understanding the unique and diverse experiences of millions of people across very different cultures.

This book has grown out of the mounting dissatisfaction that many of us scholars of disaster studies have come to feel when facing such a paradox. The paradox is that we claim to be critical while, in practice, we are probably not as critical as we claim to be or feel that we are. This dissatisfaction is probably best expressed in the manifesto *Power, Prestige and Forgotten Values*, which calls for rethinking our research agenda through more respectful, reciprocal and genuine relationships in disasters studies and which many of us co-signed in 2019 (https://www.ipetitions.com/petition/power-prestige-forgotten-values-a-disaster). This call for critical reflection and hopefully for change in our approach to studying disaster has already spurred meaningful discussions and inspired a new generation of young, creative and committed researchers to explore new pathways for studying disasters (e.g. Alburo-Cañete, 2021; Lakhina, 2020).

This book sets out to build on this momentum and provide some theoretical rationale for revisiting some of the key dimensions of disaster studies. It challenges the hegemony of Western discourses on disaster and their influence in shaping standardised disaster risk reduction policies and actions that overlook the realities of diverse localities, people and cultures. It is an endeavour of deconstruction inspired by critical social theory, in particular postcolonial and poststructuralist studies.

The sequence of working titles that this book has been through mirrors my intellectual journey since the very idea of such a text came about. Initially, the book was meant to be titled *Marginality and Disaster* to reflect the nature of my earlier work, at least that work between 2000 and 2015. Then it was to be called *Inside Out* to build on the momentum of the above-mentioned manifesto and an earlier call to move disaster studies 'inside out'. I have wandered away from both these titles for the very same reason: that is, to overcome the dialectics of centre–margin and inside–outside which I believe and argue later in the book is insufficient to open up epistemological spaces for other understandings and ways of dealing with disasters to emerge.

The Invention of Disaster may be a more provocative title. Nonetheless, I contend that it is one that allows us to capture the essence of the argument. In fact, I opened this book with an epigraph that reminds us of the different meanings of the term invention because I believe that they all somehow apply to the argument of this book. Indeed, I argue that the concept of disaster has been fabricated by the West on the basis of its own interpretation of the world, which is the realisation, the 'finding out', by Europe through the eighteenth century of the harm that natural phenomena may cause to society. As such, it reflects one cultural reality rather than a universal truth. This 'discovery of valid [in Europe] or seemingly valid [elsewhere in the world] arguments' on the reality of disaster, however, has supported a rhetoric, in Cicero's (~87 BC) terms, 'to render one's case plausible'. This case is the combined rolling-out of Western discourses and governmentality of disaster across the world as part of a broader imperialist agenda portrayed as the only path towards progress and freedom, as unveiled in the project of modernity inherited from the Age of Enlightenment.

The Invention of Disaster is a title I obviously borrow from Valentin-Yves Mudimbe's *The Invention of Africa* and Oyèrónké̩ Oyěwùmí's *The Invention of Women*, two seminal texts that have had a profound influence on this book. Both Mudimbe and Oyěwùmí, however, write from the perspective of those whose views have been neglected, if not obliterated, by Western discourses on Africa and gender. In this perspective, one may contest (and rightfully so) my own positionality and the legitimacy of my effort in driving forward a similar postcolonial agenda and endeavour of deconstruction in disaster studies. This question entails a twofold reflection that is at the core of any endeavour of deconstruction, especially in postcolonial spaces (Young, 2004): one of identity (my world/s) and one of method (my words).

The former is at the centre of the postcolonial agenda, and the question of re/presentation of non-Western identities, ontologies, epistemologies, cultures, and so on has stirred considerable attention (Chapter 1). Whether, to use Spivak's (1988, pp. 281) own words, by 'accident of birth and education' a White, middle-aged cis man who grew up in Western Europe is in a legitimate position to challenge the hegemony of Western imperialism is, in fact, a political issue as much as a matter of knowledge and identity (Chapter 3). Spivak (1987, p. 349) once elaborated that

> This is part of a much larger confusion: can men theorize feminism, can whites theorize racism, can the bourgeois theorize revolution and so on. It is when only the former groups theorize that the situation is politically intolerable. Therefore, it is crucial that members of these groups are kept vigilant about their assigned subject-positions. It is disingenuous, however, to forget that, as the collectivities implied by the second group of nouns start participating in the production of knowledge about themselves, they must have a share in some of the structures of privileges that contaminate the first group.

Of course, I have spent most of the past 24 years in the Philippines, teaching, researching and working alongside non-government and civil society organisations as well as local government agencies, advocating for the preservation of Kapampangan culture, and ultimately experiencing everyday life and what natural hazards and disasters may mean beyond Western Europe. Yes, my intellectual growth owes much, thank you, to the University of the Philippines Diliman and its network of libraries to which I will be forever grateful. But is it enough?

My position and my own Western heritage, I hope, contribute some legitimacy to my critique of this very heritage and its imposition upon other regions of the world, such as the Philippines, where I have been confronted by its legacy first-hand. My own misinterpretation and ill usage of the so-called vulnerability paradigm in my studies over the past 25 years have taught me how flawed my research was. I believe that I am not an isolated case. Similarly, my own practice of, and advocacy for, inclusion, especially through participatory mapping and other approaches to foster the participation of minority groups in reducing the risk of disaster, has made me realise how limited room there actually is for genuine action in this space. In many instances, my own practice was as skewed as my research, willingly or not.

This is why, in this book, I argue – from personal experience as much as from engagement with critical social theory – that the subalterns in postcolonial lingua should be at the core of our agenda to revisit disaster studies, whether they are local researchers or people who are dealing with what we call natural hazards and disasters in their everyday life. This is a matter of understanding local world views/senses and ways of knowing in order to address the paradox I opened this preface with (Chapter 3). This is also a matter of challenging the hegemonic neoliberal ideology that envelops the imperialist agenda that underpins both disaster studies and disaster risk reduction (Chapter 5). Both are long-term goals that extend beyond the scope of this book.

My goal in this book, as unveiled in Chapter 1, is primarily to disassemble (i.e. deconstruct) the hegemony of Eurocentric/Western ontologies and epistemologies in disaster studies and disaster risk reduction from within in order to expose their inherent flaws and irrelevance when applied outside of their cradle, namely Europe and, by extension, the rest of the Western world. I consider that this is a required preliminary step for eventually re/constructing disaster studies from multiple, subaltern perspectives from the non-Western world, the long-term goal that, I hope, may open up space for alternative approaches for addressing disaster risk, should the latter be relevant. As such, it is my contention that my position as an 'insider' to the Western world is not at odds with my argument. Spivak (1985, p. 11) once said, 'the greatest gift of deconstruction: to question the authority of the investigating subject without paralysing him, persistently transforming conditions of impossibility into possibility'.

I am conscious, however, that the very nature of this book reflects a Western approach to exposing and discussing knowledge and epistemologies, namely through words and texts articulated and structured after Western standards, which are, I will argue, inherently allegorical, in Clifford's (1986b) sense. The use of the English language similarly prolongs the hegemony of the West. Hence my use of quotes in their original language to try to preserve some of the nuances expressed by their authors, which otherwise would be smothered by a translation into English, although this still does not fully alleviate the problem since these words are from the languages of other former colonial powers. As such, my effort may be perceived as carrying some of the inherent epistemological flaws of Western scholarship that I challenge later on. Flaws that have stirred continuing and heated debates in poststructuralist and postcolonial studies since Barthes's (1968) contentious essay *La mort de l'auteur* (e.g. Foucault, 1969a; Derrida, 1981; Spivak, 1988; Bhabha, 1994).

This initial step of deconstruction from within and through text is therefore necessarily to be completed, challenged and expanded by perspectives from the non-Western world, whether expressed in words or other diverse media. As I argue later on, it is my contention that there is no single truth and knowledge about what we call natural hazards and disasters. As such, *The Invention of Disaster* is also an invitation to explore new paths. It is a hopeful perspective into the future. To follow Derrida (1987), invention, by nature, stimulates creativity, an inherent dimension of any endeavour of deconstruction.

In no way does this mean that there is no value in the heritage of the Enlightenment and its contribution to the advance of Western science and philosophy within its cradle or wherever else it makes sense within the local culture. Nor do I deny the material reality of the suffering of millions of people around the world in dealing with what we call natural hazards. Far from it. In summary, I do not suggest throwing everything out of the window, including the very concept of disaster and decades of learning in disaster studies. As Derrida once said: 'deconstruction is not negative. It's not destructive' (Derrida and Mortley, 1991, p. 96). The point is, once again, to question and challenge the hegemonic rolling-out of both Western discourses on disaster and the imperialist *dispositif* of disaster risk reduction in places where they do not make sense.

I offer the following collection of essays as a potential springboard for envisioning a more diverse future. Some of these essays were published throughout the past decade but have been significantly revisited for this book. They are completed by six novel contributions (Chapters 1, 5, 6, 7, 9 and 10) that expand and update some dimensions of my argument. These chapters have benefited from the thoughts and feedback of colleagues and friends who deserve due credit for their continuing support, constructive criticism and encouragements in pulling this book together.

I am particularly thankful to both Ksenia Chmutina (Loughborough University) and Ben Wisner (University College London), who have provided extensive and continuing guidance throughout my writing journey. I am grateful to Virginia Garcia-Acosta (Centro de Investigaciones y

Estudios Superiores en Antropologia Social), Emmanuel Raju (University of Copenhagen), Maureen Fordham (University College London), Ken Hewitt (Wilfrid Laurier University), James Lewis (Datum International), Mihir Bhatt (All India Disaster Mitigation Institute), Christopher Gomez (Kobe University), Jake Rom Cadag (University of the Philippines), Emmanuel Luna (University of the Philippines), Soledad Natalia Dalisay (University of the Philippines), Raymund Narag (Southern Illinois University), Ilan Kelman (University College London), Maria Fellizar-Cagay (Center for Disaster Preparedness), Zenaida Delica-Willison (Center for Disaster Preparedness), Rohit Jigyasu (International Centre for the Study of the Preservation and Restoration of Cultural Property), Lori Peek (University of Colorado Boulder), Sylvie Devèze (Université Paul Valéry – Montpellier III), Claude Gilbert (Centre National de la Recherche Scientifique), Katherine Hore (The University of Auckland), Alice McSherry (The University of Auckland), Louise Baumann (The University of Auckland), Anthony Gampell (The University of Auckland), Tanay Amirapu (The University of Auckland), Martin Joe (The University of Auckland) and Loïc Le Dé (Auckland University of Technology) for their invaluable suggestions on early drafts of this book and/or multiple discussions that have informed some dimensions of my argument. The expert editing skills of Peta Stavelli have also been greatly appreciated in finalising the manuscript.

In addition, I would like to express my sincere words of gratitude to the colleagues with whom some of the original essays revisited here were authored. These acknowledgements extend to the editors of books and journals (Noel Castree, Sara Pantuliano, Dewald van Niekerk and Lisa Schipper) who have allowed some sections of these essays to be republished here. These original essays include the following:

Gaillard J.C. (2017) Natural hazards and disasters. In Richardson D., Castree N., Goodchild M.F., Kobayashi A.L., Liu W., Marston R. (eds.) *The international encyclopedia of geography: people, the earth, environment and technology. Association of American Geographers.* Chichester, UK: Wiley-Blackwell, 1–15. (Chapter 2)

Gaillard J.C. (2019) Disaster studies inside out. *Disasters* 43(S1): S7–S17. (Chapter 3)

Gaillard J.C., Gomez C. (2015) Post-disaster research: is there gold worth the rush? *Jàmbá: Journal of Disaster Risk Studies* 7(1): http://jamba.org.za/index.php/jamba/article/view/120. (Chapter 3)

Gaillard J.C., Jigyasu R. (2016) *Measurement and evidence: whose resilience for whom?* Working Paper Series No. 11, Resilience Development Initiative, Bandung. (Chapter 4)

Gaillard J.C. (2012) The climate gap. *Climate and Development* 4(4): 261–264. (Chapter 6)

Gaillard J.C., Sanz K., Balgos B.C., Dalisay S.N.M., Gorman-Murray A., Smith F., Toelupe V. (2017) Beyond men and women: a critical perspective on gender and disaster. *Disasters* 41(3): 429–447. (Chapter 8)

I would like to acknowledge that the ideas expressed in this book and its critical and probably challenging argument as well as its flaws and limitations are mine and only mine. Please be kind and constructive readers.

JC Gaillard
Tāmaki Makaurau, Aotearoa
February 2021

1 What is a disaster?

What is a disaster? This question has puzzled generations of researchers across fields and disciplines, none of whom has been able to offer any consensual definition (Quarantelli, 1998; Perry and Quarantelli, 2005). This book is not yet another attempt at answering this question, which is, to our view, an aporia. All definitions in a broad and eclectic field of scholarship such as disaster studies will always reflect disciplinary assumptions and objectives that researchers carry with them when engaging with the concept of disaster. On the other hand, nor does this book endeavour to throw the concept of disaster out of the window. As Derrida (1967a, p. 25) once said, '*nous devons d'autant moins renoncer à ces concepts qui nous sont indispensables pour ébranler aujourd'hui l'héritage dont ils font partie*'[1].

Rather, this book questions how and why we ask what a disaster is in the first place. It also explores what such questioning entails in terms of scholarship and the approaches we design in attempting to reduce disaster risk. As such, this book is inevitably an endeavour of deconstruction, as in Derrida's (2004a, p. 1100) own words:

> c'est une pensée de l'origine et des limites de la question "qu'est-ce que?...", la question qui domine toute l'histoire de la philosophie. Chaque fois que l'on essaie de penser la possibilité du "qu'est-ce que?...", de poser une question sur cette forme de question, ou de s'interroger sur la nécessité de ce langage dans une certaine langue, une certaine tradition, etc., ce qu'on fait à ce moment-là ne se prête que jusqu'à un certain point à la question "qu'est-ce que?". C'est ça, la différence de la déconstruction. Elle est en effet une interrogation sur tout ce qui est plus qu'une interrogation.[2]

Our interrogation focuses on the ontological, epistemological and ideological foundations that underpin disaster studies and disaster risk reduction as entangled fields of scholarship, policy and action that have gained considerable and worldwide traction over the past hundred years, since at least Samuel Prince's PhD 1920 thesis on the Halifax disaster. We particularly explore the universal relevance and influence of a concept of Latin etymology and how this concept, alongside some of its cognates (hazard,

DOI: 10.4324/9781315752167-1

vulnerability, resilience, capacities, etc.), has informed policies and actions in places where they do not translate or even make sense. Therefore, asking what a disaster is is more than a definitional question. It requires the exploration of knowledge and power structures in disaster studies and disaster risk reduction. What and whom do we study? And why? Who does the studying? Where? And from which perspectives? How does knowledge generated by researchers sustain policies and actions? Ultimately, how do disaster scholarship and disaster risk reduction initiatives contribute to sustain a broader ideology that shapes the world we know today?

As such, this book endorses Foucault's (1969b) call to challenge what is taken for granted in our interpretations of the world and it unpacks processes that have led to these interpretations. In disaster studies, answering this call entails confronting the Eurocentric/Western ontological and epistemological heritage of the Enlightenment and how it has sustained the imperialist and ethnocentric ideology promoted by the West. Our argument is that the hegemony of Western knowledge in disaster studies supports normative and standardised disaster risk reduction policies and actions, which in many instances fail to consider the diverse realities of very different local contexts around the world.

Power and knowledge

This book is therefore about power and knowledge in disaster. Our approach to power owes a debt to both Foucault (1975, 1976) and Gramsci (1971), although we recognise that there are other, dissonant perspectives (see Lukes, 2005 and Morriss, 2002 for recent overviews). To both Foucault and Gramsci, power is not a thing that emanates from particular institutions, individuals or sets of actions at a macro or superstructural level that is imposed on others. Power is about unequal relationships that allow some institutions and individuals to guide and control the behaviours of others (Foucault, 1976, 1982). As such, power is embedded in time and space, within all dimensions of society and the everyday lives of people from the state level as much as within families. It is therefore concomitant to other forms of economic, political and interpersonal relationships that bind the social fabric.

The exercise of power may be coercive and tangible or more subtle, invisible, and based on consent. Traditional views of power have long considered power relations as direct, from A over B (Bachrach and Baratz, 1962; Dahl, 1957). Yet the exercise of power is also indirect, multi-nodal, fluid, mobile and engrained within policies, regulations, institutions, literature and the arts as well as architecture and infrastructure such as schools, workshops, hospitals, prisons and museums that shape the everyday lives of people and subjectify them to the will and control of those in power (Althusser, 1975; Foucault, 1982). In a seminal definition, Foucault (1976, pp. 122–123) summarises:

> le pouvoir est partout; ce n'est pas qu'il englobe tout, c'est qu'il vient de partout. (…) Ce n'est pas une institution, et ce n'est pas une

structure, ce n'est pas une certaine puissance dont certains seraient dotés: c'est le nom qu'on prête à une situation stratégique complexe dans une société donnée[3].

Power is also internalised and mediated by the psyche (Butler, 1997). As such, power is exercised beyond the realm of the sole political. Gramsci (1930, 1971) actually locates power in the cultural sphere and associates its exercise with intellectuals such as teachers, religious figures and journalists who critically contribute to spreading dominant ideas. For Gramsci and Althusser (1975), the exercise of power as a mechanism of control and subjection is therefore inherently ideological.

Exercising power has to be justified by a dominant form of knowledge that is considered and presented as truth by those whose aim is to guide and control the behaviour of others (Foucault, 1975, 1976). Exercising power requires a fine knowledge and close monitoring of those whose behaviour is to be guided and controlled. In Western societies, power relies upon scientific knowledge as an expression of truth or reason and state or institutional knowledge (i.e. statistics as a means of control and normalisation of people's behaviour). Power and knowledge therefore are intimately linked to each other in all dimensions of the social fabric and everyday life. They are 'synonymous', to quote Horkheimer and Adorno (1947).

Power and knowledge come together through discourses (Foucault, 1976). Foucault (1969b, p. 148) defines a discourse as '*ensemble des énoncés qui relèvent d'un même système de formation*', which entails that they have emerged through the same frame at a given time and are structured through common concepts and strategies to serve a similar function. Discourses therefore are performative in the sense that they enact power (Butler, 1993). Their reiterative nature contributes to producing the effects that they name, regulate and constrain as well as to imposing dominant forms of knowledge and the interpretation of objects and phenomena through multiple channels, policies, institutions, media, forms of architecture, and so on. As such, discourses contribute to the social existence and significance of these particular objects and phenomena, including those we call natural hazards and disasters. Laclau and Mouffe (2001, p. 108) famously reflected on the relevance of discourses in the social construction of earthquakes:

> An earthquake or the falling of a brick is an event that certainly exists, in the sense that it occurs here and now, independently of my will. But whether their specificity as objects is constructed in terms of 'natural phenomena' or 'expressions of the wrath of God', depends upon the structuring of a discursive field. What is denied is not that such objects exist externally to thought, but the rather different assertion that they could constitute themselves as objects outside any discursive condition of emergence.

It is important to note, though, that it is not our intention to deny the material existence of natural phenomena such as earthquakes, volcanic eruptions, cyclones or floods. Our point is that it is through particular discourses, and the underpinning influence of specific forms of knowledge, that, in some societies, these natural phenomena are interpreted as hazards and elsewhere as resources or retributions for whatever misconduct. It is therefore not the concepts, per se, that stir our attention in this book, but rather the interpretation and translation of particular objects and phenomena into these concepts and how these concepts hence reflect the view of those who use them more than that of those who are directly confronted by them, if the former and the latter are not the same individuals.

In Foucault's approach to power, scientific knowledge and associated narratives, discourses, policies, regulations, institutions as well as forms of architecture and infrastructure are gathered within *dispositifs*. Foucault defines a *dispositif* as follows:

> un ensemble résolument hétérogène, comportant des discours, des institutions, des aménagements architecturaux, des décisions règlementaires, des énoncés scientifiques, des propositions philosophiques, bref, du dit, aussi bien que du non-dit, voilà les éléments du dispositif. Le dispositif lui-même, c'est le réseau qu'on peut établir entre ces éléments. (…) une sorte – disons – de formation, qui, à un moment historique donné, a eu pour fonction majeure de répondre à une urgence. Le dispositif a donc une fonction stratégique dominante.[4]
>
> (Foucault et al., 1977, p. 63)

Power is, however, not all about control and domination. The exercise of power also triggers resistance and may be enabling for some actors and organisations. Resistance may take multiple forms, including *contre conduites* (Foucault, 2004a), war of position (Gramsci, 1971), everyday techniques and means of class struggle (Scott, 1985, 1990), creative tactics to reinterpret and reappropriate strategies and methods of control (de Certeau, 1980), counter-insurgency strategies (Hardt and Negri, 2004), civil society networks and advocacy (e.g. a genuine approach to participation) (Bello, 2001; Norton and Gibson, 2019), hybridization of knowledge and practices (Bhabha, 1994) and even collaboration (Guha, 1997).

Notwithstanding differences in their expression and materialisation, these diverse forms of resistance are all embedded within relationships of power yet not as the other end of a binary relationship (Foucault, 1976). Because power is everywhere, often invisible and embedded within the everyday social fabric, so are the different forms of resistance. Relations of power and resistance thus appear as a field of diffuse forces exercised from both the top down and the bottom up.

Unpacking the intrinsic relationship between power and knowledge in disaster requires us to focus on how knowledge on disaster is created and embedded within particular discourses that are shared and imposed through a sophisticated *dispositif*. It necessitates an understanding of how this knowledge and the discourses it underpins sustain strategies of disaster risk reduction that are designed by certain actors and imposed upon, or resisted by, people who are considered at risk and whose responses and behaviours are to be guided or controlled (or both). Our objective is therefore epistemological in nature but necessarily requires us to explore the ontological assumptions that underpin our ways of knowing and the ideology the latter supports.

Understanding disaster and the Enlightenment legacy

At the crux of this exploration is the ontological assumption that disasters, at least those associated with natural phenomena, sit within the nature–culture binary or, in the disaster studies jargon, between hazard and vulnerability. Indeed, it is widely acknowledged, if not taken for granted, that the concept of disaster captures harm and damage caused when a hazardous phenomenon affects vulnerable people and their livelihoods. This ontological assumption can be traced back to the eighteenth century, as emphasised in the famous dialogue between Voltaire and Rousseau following the 1755 Lisbon earthquake, eventually dubbed the first modern disaster (Dynes, 2000).

As such, modern understandings of disaster are firmly grounded in a broader legacy of the Enlightenment, or Age of Reason that Kant (1784) famously summarised as follows:

> Aufklärung ist der Ausgang des Menschen aus seiner selbstverschuldeten Unmündigkeit. Unmündigkeit ist das Unvermögen, sich seines Verstandes ohne Leitung eines anderen zu bedienen. Selbstverschuldet ist diese Unmündigkeit, wenn die Ursache derselben nicht am Mangel des Verstandes, sondern der Entschließung und des Mutes liegt, sich seiner ohne Leitung eines andern zu bedienen. Sapere aude! Habe Mut, dich deines eigenen Verstandes zu bedienen! ist also der Wahlspruch der Aufklärung.[5]

Ausgang here includes freedom, or emancipation, from the hazards of nature that the 1755 earthquake made pressing. Rousseau's (1967) interrogation on the causes of the disaster indeed marks the transition from a representation of disasters through the lens of natural phenomena and their previously evident impact – such as in Defoe's (1704) *The Storm* or in the different entries listed in Chambers's *Cyclopaedia* and Diderot and D'Alembert's *Encyclopédie* – to a questioning of explanations in their historical dimension about the development of the city of Lisbon in a way that made it vulnerable to an earthquake.

In fact, Kant himself, who once blamed God's will (1755), would eventually revisit his view on the impact of earthquakes and other natural phenomena to state that '*er der Vorsehung wegen der Übel, die ihn drücken, keine Schuld geben müsse (...) sich also von allen Übeln, die aus dem Mißbrauche seiner Vernunft entspringen, die Schuld gänzlich selbst beizumessen habe*'[6] (Kant, 1786,

p. 123). This increasing concern for the causes of disasters, especially in relation to people and their agency, constitutes an excellent example of the quest for deeper analysis, life and historicity that Foucault (1966) associates with the modern *épistémè* of the West that has resulted from Enlightenment thinking. This thinking was assumed by the local scientists and philosophers of the time to establish Europe's supremacy over the world, as in the Chevalier de Jaucourt's (1756, T6, p. 212) entry in the *Encyclopédie*: Europe is '*la plus considérable de toutes* ['part of the world'] *par son commerce, par sa navigation, par sa fertilité, par les lumieres & l'industrie de ses peuples, par la connoissance des Arts, des Sciences, des Métiers*'[7].

Locating our argument within this ontological heritage is essential because the nature/hazard–culture/vulnerability binary, celebrated in the famous mnemonic 'disaster = hazard × vulnerability' (or any iterations of this), has since polarised disaster studies (as discussed in Chapter 2). A prominent trait of these studies is that they have long engaged in a typically Hegelian dialogue between the relative importance of nature and culture/society in explaining the occurrence of disasters and the response of people to them. The two main paradigms available in disaster studies, as well as other cognate approaches, including those positioning themselves right at the interface, are all situated within the space offered by the nature/hazard–culture/vulnerability binary. Hence, the seemingly irreconcilable differences between the mainly positivist hazard paradigm and the originally constructivist vulnerability paradigm have steered much debate, but such debate has remained bounded by the narrow scope provided by Eurocentric concepts such as hazards and vulnerability as much as it has been supported by methodologies similarly inherited from the Enlightenment.

If the dialectical debate between the hazard and vulnerability paradigms has clearly marked a displacement of the centre of attention from nature/hazard to culture/vulnerability, it has thus in no way led to a reconsideration of the underpinning binary. This binary is central to the modern *épistémè* of the West inherited from the Enlightenment (Castree, 2014; Horkheimer and Adorno, 1947; Smith, 2008). In this perspective, one may argue that such a binary view of disasters, and of the world at large, reflects a largely structuralist heritage, which Foucault (1966, p. 221) recognised as '*la conscience éveillée et inquiète du savoir moderne*'[8]. Scholarship on disaster has been fragmented into the separate study of different hazards, split across a number of sciences, and different dimensions of personal and organisational responses through perspectives, especially Marxist, that depart from structuralism. But in the end, all of these sit within the reductionist view that disasters mirror the nature/hazard–culture/vulnerability binary, which is structuralist in Lévi-Strauss's (1962) terms.

The largely dialectic nature of the paradigms that have informed disaster studies to date, and their extension into discourses that inform disaster risk reduction, is not only a matter of concepts and paradigms stuck within the nature–culture binary. We show in Chapter 3 how it has restricted our understanding of disasters in such a narrow frame that has excluded the possible emergence of other, radically different perspectives. According to Derrida (1967b, p. 409), centres within a structure are inhibiting:

> ce centre avait pour fonction non seulement d'orienter et d'équilibrer, d'organiser la structure – on ne peut en effet penser une structure inorganisée – mais de faire surtout que le principe d'organisation de la structure limite ce que nous pourrions appeler le jeu de la structure[9].

As such, the very binary and the different paradigms it has sustained have been considered the only relevant truth, one that is dictated by reason and Western science. As a result, studies of disasters remain firmly grounded in the modern *épistémè* of the West inherited from the Enlightenment. Such a heritage has come down a long way and trickled through the concepts we coin, the theories we develop, the methodologies we resort to and even the ways we share our findings (see Chapters 2, 3 and 4). The legacy of the nature/hazard–culture/vulnerability binary further extends to the policies and actions designed to address disaster risk.

Disaster risk reduction and the project of modernity

The modern *épistémè* of the West, out of which Habermas (1985) carved the project of modernity, stemmed directly from the Enlightenment. Enhancing our understanding of nature through science to eventually prevent its hazards in a way that fosters progress and frees people from the adverse effects of nature was one central dimension of Kant's call for *Ausgang*. Harvey (1990, p. 12) describes the project of modernity in the following terms:

> That project amounted to an extraordinary intellectual effort on the part of the Enlightenment thinkers "to develop objective science, universal morality and law, and autonomous art according to their inner logic" [quoting Habermas]. The idea was to use the accumulation of knowledge generated by many individuals working freely and creatively for the pursuit of human emancipation and the enrichment of daily life. The scientific domination of nature promised freedom from scarcity, want, and the arbitrariness of natural calamity. The development of rational forms of social organization and rational modes of thought promised liberation from the irrationalities of myth, religion, superstition, release from the arbitrary use of power as well as from the dark side of our own human natures. Only through such a project could the universal, eternal, and the immutable qualities of all of humanity be revealed.

Advancing this agenda has required new forms of government, in its literal sense: that is, to govern someone's behaviours through sophisticated and subtle strategies and techniques of control and normalisation. This exercise of power could not appear coercive and, as such, had to be accepted and embedded within the everyday lives of those whose behaviour was to be controlled. Foucault (2004a) calls this art of government *governmentality*. This includes both these power relations and the strategies and techniques that underpin them. These strategies and techniques are articulated from three perspectives that come together as a triangle: the expression of sovereignty by the government, the imposition of discipline upon people whose lives are guided and controlled, and governmental management or the institutions and techniques that allow for sovereignty and discipline to be exercised (Foucault, 2004a).

Disasters, and most recently climate change (Chapter 6), have provided very fertile grounds for deploying a specific *dispositif* within such form of governmentality to achieve the goals of the project of modernity. The alleged *extra*-ordinary dimension of disasters indeed offers a unique opportunity for governments and institutions of power to flex their muscle and affirm their sovereignty upon their territory. The extensive damage that they sometimes cause further provides planners and architects with the blank slate needed to rethink the affected regions, especially cities, from a new and rationalised perspective which favours social control. To use Harvey's (1990) words, disasters are crucibles of 'creative destruction'.

Disaster risk reduction, in the broadest sense, which includes response and recovery as well as climate change adaptation, offers additional opportunities to justify and implement normative and standardised strategies for guiding the behaviour of people at risk as well as that of those who are affected. In Chapters 5, 6 and 7, we discuss how both the hazard and vulnerability paradigms in turn have sustained their own sets of normative and standardised initiatives firmly grounded within the Enlightenment commitment to control both nature and society. The former has paved the way for hazard-centred, technocratic and command-and-control strategies that have built upon scientific knowledge. The latter's emphasis on inclusion, participation and 'community-based/led/managed' actions has provided seemingly pluralistic and context-specific opportunities but ultimately packaged these within a similarly normative and often romanticised view of 'communities' as homogenous and place-based social entities (Cannon, 2014).

These normative and standardised strategies trickle through multiple pathways that include explicit civil defence/protection strategies which mimic the command-and-control approach of military organisations or more subtle approaches through schools and civil society organisations, as in the case of 'community-based/led/managed' strategies. Both sets of initiatives rely heavily upon rational and forward-looking approaches, so that the concept of planning is a crucial component of disaster risk reduction. There are disaster risk reduction plans, emergency management plans, recovery plans, and so on, at all levels of policy and action from international to local/'community' through to national, all of which are geared to lessen damage should a potentially harmful natural phenomenon occur.

These plans are designed to order things, whether these are land use and infrastructure, administration, procedures or behaviours, in a way that is rational and rigorous and to the benefit of a progressive society as a whole (Harvey, 1985; Gunder, 2003). Therefore, as much as planning is crucial to disaster risk reduction, disaster risk reduction is essential to implementing the project of modernity. As a result, planning is widely criticised for being the ultimate instrument of social control and oppression designed to advance the project of modernity (Yiftachel, 1998; Flyvbjerg, 1998). As Gunder (2003, p. 238) notes,

> [p]lanning is not just a product of the enlightenment, or modernity, it is fundamentally a child of their foundational metaphysical 'essence'— Western rationality. As such, it is premised on privileging one side of a binary opposition (positive over negative, presence over lack, mind over body, right over left, remedy over poison) and has no scope for ambiguity.

This one-sided nature of planning raises the question of whose rationality is being heard and listened to when devising plans and associated policies, strategies, and so on and whose lives and livelihoods planning is meant to enhance (Harvey, 1985). Since planning is by nature rigid and technocratic, it is well known for being exclusive rather than inclusive. When it claims to be inclusive, such as through the rhetoric of participation – and 'community-based/led/managed' disaster risk reduction is an excellent example of this (see Chapter 7) – it is often skewed to the benefit of a powerful few and to the detriment of most people who deal with what we call natural hazards and disasters (Cooke and Kothari, 2001). This is particularly true in societies, beyond the West, where the very idea of planning does not make sense, such as in the case of Kiribati, which we explore in Chapter 6 (Chambers, 1974; Cooke and Kothari, 2001).

The project of modernity indeed is associated with colonial expansion and economic globalisation supported by the alleged desire to bring the benefits of Western science and progress to societies beyond Europe (Quijano, 1992; Blaut, 1993; Mignolo, 2011). It is in this context that Western discourses in disaster studies and associated policies and actions to reduce disaster risk have been rolled out around the world, beyond their European cradle, as the only available options to address disasters.

The hegemony of Western discourses

Our framing of hegemony owes a debt to both Gramsci and Said. Gramsci's (1971) approach to hegemony, which is based on consent rather than force and coercion in exercising power, provides a space to understand how Western discourses on disasters are taken for '*common sense*' and considered the only options to both understand and address disaster risk, wherever we are in the world. In this regard, the hegemony of Western discourses that frame disaster studies and disaster risk reduction constitutes one dimension of Western imperialism over the rest of the world (Said, 1994). We assume that disasters are universally the consequence of the unfortunate interaction between nature and culture and that science and technology can address this as part of the Enlightenment's quest for modernity and control over nature (Hewitt, 1983).

Understanding the hegemonic nature of Western discourses on disaster, and the imperialist ideology they sustain, therefore requires us to unpack the intimate relationships between power and knowledge within the space of Western science and policy but also through the broader relationships of domination that exist between the West and the rest of the world, as was brilliantly summarised by Said (1978, p. 5) in the context of what he calls Orientalism:

> ideas, cultures, and histories cannot seriously be understood or studied without their force, or more precisely their configurations of power, also being studied. To believe that the Orient was created – or, as I call it, "Orientalized" – and to believe that such things happen simply as a necessity of the imagination, is to be disingenuous. The relationship between Occident and Orient is a relationship of power, of domination, of varying degrees of a complex hegemony.

Locating disaster studies and disaster risk reduction within the broader imperialist ideology of Europe and its allies is essential as ideology sustains hegemony (Mudimbe, 1988; Althusser, 2018), especially within colonial or post-colonial contexts (JanMohamed, 1985; Shohat, 1992). The hegemony of Western discourses in disaster studies and disaster risk reduction thus mirrors a form of ethnocentrism. In Said's words (1985, p. 100), 'the production of knowledge, or information, of media images, is unevenly distributed: its locus and the centers of its greatest force are located in what, on both sides of the divide, has been polemically called the metropolitan West'. In this perspective, 'the analysis of the dialectics of the centre and the margin can thus operate geographically as well as conceptually, articulating the power relationships between the metropolitan and the colonial cultures at their geographical peripheries' (Young, 2004, p. 50). This point is further explored in Chapter 3.

The hegemony of Western discourses on disaster has built upon broader mechanisms of control and domination associated with colonial and neocolonial strategies that Memmi (1957, p. 90) summarises as follows:

> un effort constant du colonialiste consiste à expliquer, justifier et maintenir par le verbe comme par la conduite, la place et le sort du colonisé, son partenaire dans le drame colonial. C'est-à-dire, en définitive, à expliquer, justifier et maintenir le système colonial, et donc sa propre place. Or l'analyse de l'attitude raciste y révèle trois éléments importants:

1. Découvrir et mettre en évidence les différences entre colonisateur et colonisé.
2. Valoriser ces différences, au profit du colonisateur et au détriment du colonisé.
3. Porter ces différences à l'absolu en affirmant qu'elles sont défini-tives, et en agissant pour qu'elles le deviennent.[10]

Understandings of disasters and strategies to reduce disaster risk very much build upon such fabricated divides between, on the one hand, a 'safe' West that suffers fewer human losses and has accumulated sufficient knowledge of disaster risk and developed appropriate strategies to address that risk and, on the other hand, a 'dangerous' rest of the world that lacks both an understanding of the problems and the resources to address them (Hewitt, 1995a; Bankoff, 2001). It is this process that emphasises an alleged nexus between modernity and rationality that Peruvian sociologist Aníbal Quijano (1992) coined *coloniality*. Indeed, it is through colonisation and the alleged/fabricated differences between the coloniser who knows best and the colonised who knows less that Europe has imposed its own understand-ing of disasters and ways of dealing with them upon the rest of the world. In Foucault's (1977a, p. 60) own words:

A partir del siglo XIX, hay que decir sin duda que los esquemas de pen-samiento, las formas políticas, los mecanismos económicos fundamen-tales que eran los de Occidente se universalizaron por la violencia de la colonización, o, bueno, digamos que la mayoría de las veces cobra-ron de hecho dimensiones universales. Y eso es lo que entiendo por Occidente, esa suerte de pequeña porción del mundo cuyo extraño y violento destino fue imponer finalmente sus maneras de ver, pensar, decir y hacer al mundo entero[11].

This perspective is sustained by data and statistics – an essential foundation of the Western modern *épistémè* (Foucault 2004a; Horkheimer and Adorno 1947) that hides local realities. This data nonetheless is compiled in inter-national databases that serve as the ultimate truth and basis for both schol-arship and international policy agendas (Chapters 4 and 5).

This fabricated divide between the West and the rest of the world sustains the alleged prominent position of Europe and its allies in dealing with disasters, thus reinforcing the primacy of Western ontologies and episte-mologies. As a result, in line with Memmi's third point, Western ontologies and epistemologies become 'common sense'. They are taken for granted and considered the single, absolute truth by everyone, both in the West and beyond. Memmi (1957, p. 120) adds: '*la sclérose de la société colonisée est donc la conséquence de deux processus de signes contraires: un enkystement né de l'intérieur, un corset imposé de l'extérieur*'[12].

The mechanisms of diffusion of this unconscious consent to Western reason and culture, in disaster studies and beyond (Fanon, 1952), are multiple and engrained within the lives and roles of those who deal with what we call disasters. They are taught to us from a very young age at school. They are channelled by media reports on disasters. They are requirements in regard to how academic publications have to be framed and structured. They ultimately provide templates for the disaster risk reduction policies and actions that we tackle in Chapters 5 and 6. As Peet (2007, p. 56) summarises, 'Western rationality is the hegemonic global form of deep and careful thought applied to scientific, technical and policy matters'.

Disaster risk reduction strategies indeed build upon the alleged truth provided by Western discourses in disaster studies. The progressive shift from a hazard-driven perspective to the so-called 'community-based/led/ managed' approach observed over the past three decades, and critiqued in Chapters 5 and 7, has allegedly led to a radical transformation of how we tackle disasters. This shift is yet no more than the dialectical displacement of the centre within the ontological assumption that disasters sit at the interface between nature–hazard and culture–vulnerability. As such, it perpetuates rather than challenges the hegemony of Western discourses in disaster studies and their extension to disaster risk reduction. Bankoff (2019, p. 234) summarises:

> there seems to be no escape from remaking the world again and again in accordance with a particular image. No rival discourse seems ready yet to challenge Western hegemony in the language and metaphor of international governance and development policy.

The continuing Western hegemony and imperialist nature of disaster risk reduction strategies are evident in the persistent influence of Western countries in producing international policy guidelines and agreements, from the International Decade for Natural Disaster Reduction in the 1990s to the Hyogo Framework for Action in the 2000s and the Sendai Framework for Disaster Risk Reduction in the 2010s. All of these international frameworks have encouraged the transfer of knowledge, experience, technology and other resources from the West, which allegedly knows how to deal with disasters, to the rest of the world, which is considered unable to deal with such threats alone (Bankoff, 2001). In line with such recommendations, donor agencies from Western countries and international organisations controlled by the same Western countries impose their agenda upon recipient countries and organisations through multiple strings and accountability mechanisms that encourage the latter to design their policies and actions after normative templates which mimic and extend the relevance of Foucault's triangle of governmentality to the international scene.

The hegemonic heritage of Western discourses, which Foucault (1997, p. 9) called '*tyrannie des discours englobants*', and its lingering effect on how we have been designing policies and actions to reduce disaster risk are increasingly causing discomfort amongst researchers and practitioners (Hewitt, 1995a; Bankoff, 2001, 2019). The theoretical supremacy of

Western reason that Husserl (1954) argued in his famous 1935 Vienna lecture is now widely challenged, and the rise of poststructuralist thoughts in the 1970s has confronted the relevance of any *métarécit* that is to be applied to any contexts around the world (Lyotard, 1979). For Derrida (1972, p. 254), the hegemony of Western ontologies and epistemologies shapes a 'white mythology': '*l'homme blanc prend sa propre mythologie, l'indo-européenne, son logos, c'est-à-dire le mythos de son idiome, pour la forme universelle de ce qu'il doit vouloir encore appeler la Raison*'[13].

The hegemony of Western discourses on disaster obviously fails to recognise the diversity of situations around the world – the diversity of geographical contexts, cultures and societies as much as the diverse realities and priorities of the everyday lives of people. Mbembe (1990, p. 9) brilliantly summarised this skewed perspective in the context of Africa:

> ce discours de platitudes a conduit à de fausses oppositions en vertu desquelles les commentateurs (occidentaux ou autochtones) s'estiment obligés de recourir, soit à des clichés sur les supposées spécificités culturelles (oubliant qu'elles peuvent n'être qu'un langage au sujet de très prosaïques modes de contrôle social), soit à un jugement de l'"Autre" à l'aune des valeurs dites occidentales que l'on projette sur l'objet de la recherche, se dispensant ainsi de rendre compte des équilibres effectifs produits par des cultures et des histoires différentes[14].

In fact, the hegemony of Western discourses on disaster has largely prevented us listening to or even hearing the diverse voices of those affected by disasters in a way that is meaningful to them, not just to the researchers whose agenda is driven by Western concepts, methodologies and funding schemes.

On representation and asymmetric ignorance

The question of representation is essential to understanding the hegemony of Western discourses in disaster studies and their extension to disaster risk reduction. It is twofold (Spivak, 1988): representation, as in speaking/writing for or on behalf of someone else, and re-presentation, as the actual exhibition or arrangement of ideas, as in a work of art or in a book. The two forms of representation are intimately linked in disaster studies and disaster risk reduction.

Who speaks (on whose behalf), who writes and ultimately defines what a disaster is, what it entails and what should be done to reduce risk contribute to outlining specific discourses. In discourses, knowledge, representation and reality conflate. One must conform to these discourses to communicate, to be heard, to be understood and ultimately to inform policy and actions (Foucault, 1969b). In disaster studies, dominant discourses, associated with the hazard and vulnerability paradigms, reflect Western representations of disaster, as discussed in Chapters 2 and 3. These may be valid when studying and addressing disaster risk in Europe. Problems arise when such representations are assumed to be appropriate elsewhere in the world too, across diverse cultures and societies. Then, representation may turn into 'fantasy', an 'invention' or a 'fabulation' rather than reality (Said, 1978; Mudimbe, 1988; Young, 1995a; Mbembe, 2013).

The representation of Latin America, Africa, Asia and the Pacific, or what is now known as the Global South, as 'dangerous' and vulnerable to disasters is one of these fantasies fabricated through the lens of Western discourses of disasters and it is supported by allegedly comprehensive (but, in reality, skewed) data that reflects what the West thinks a disaster is. This fantasised representation of the non-Western world very much fits within and expands what Said (1978, p. 12) called Orientalism. In his own words:

> Orientalism is not a mere political subject matter or field that is reflected passively by culture, scholarship, or institutions; nor is it a large and diffuse collection of texts about the Orient; nor is it representative and expressive of some nefarious "Western" imperialist plot to hold down the "Oriental" world. It is rather a distribution of geopolitical awareness into aesthetic, scholarly, economic, sociological, historical, and philological texts; it is an elaboration not only of a basic geographical distinction (the world is made up of two unequal halves, Orient and Occident) but also of a whole series of "interests" which, by such means as scholarly discovery, philological reconstruction, psychological analysis, landscape and sociological description, it not only creates but also maintains; it is, rather than expresses, a certain will or intention to understand, in some cases to control, manipulate, even to incorporate, what is a manifestly different (or alternative and novel) world; it is, above all, a discourse that is by no means in direct, corresponding relationship with political power in the raw, but rather is produced and exists in an uneven exchange with various kinds of power, shaped to a degree by the exchange with power political (as with a colonial or imperial establishment), power intellectual (as with reigning sciences like comparative linguistics or anatomy, or any of the modern policy sciences), power cultural (as with orthodoxies and canons of taste, texts, values), power moral (as with ideas about what "we" do and what "they" cannot do or understand as "we" do). Indeed, my real argument is that Orientalism is—and does not simply represent—a considerable dimension of modern political-intellectual culture, and as such has less to do with the Orient than it does with "our" world.

It is therefore essential to reflect upon who speaks on behalf of whom or who represents the 'other/s', the subalterns in Gramsci's (1971) vocabulary, in disaster studies and disaster risk reduction. Representation is to be considered on two levels: one that allows researchers to speak on behalf of the 'researched' and one that mirrors power relations among researchers. The former opens up questions, explored in Chapter 4, about whether there is any space for the researched to speak up and have their voice at least heard or at best considered in academic knowledge, policy and actions (Spivak, 1987). The latter raises the question of awareness and consciousness among those who are dominated and whose behaviour is guided and controlled. Consciousness is, in fact, essential to Gramsci's (1971) 'war of position' to overturn the hegemony of a dominant or ruling ideology (Chapter 10).

We show in Chapter 3 that disaster studies, as a field of scholarship, is dominated by researchers from Western countries who conduct investigations, both in their homeland and overseas, especially in Africa, Asia and the Pacific, taking advantage of available resources and incentives provided by their own institutions and the international academic industry. These researchers further build upon the alleged primacy provided by the dominant discourses on disaster that are considered the ultimate truth. As such, Western theories, concepts, methodologies and publications are assumed to be the only available resources to conduct research on disaster anywhere in the world. As Marx (1852) famously stated in *Der Achtzehnte Brumaire des Louis Bonaparte*:

> sie können sich nicht vertreten, sie müssen vertreten werden. Ihr Vertreter muß zugleich als ihr Herr, als eine Autorität über ihnen erscheinen, als eine unumschränkte Regierungsgewalt, die sie vor den andern Klassen beschützt und ihnen von oben Regen und Sonnenschein schickt[15].

This skewed representation of disaster beyond the West is organically tied to colonial heritages (Said, 1978; JanMohamed, 1985). Western scholars have been able to impose their own representation of disasters, assess their impact and explore their causes against their own criteria and expectations because European countries and their allies have had the opportunity of occupying and governing territories beyond the realm of Europe. Symmetrical opportunities have never happened for scholars beyond the West. Furthermore, because disaster studies and disaster risk reduction are dominated by Western scholars and Western organisations, the audience for any forms of knowledge produced through scholarship and policy/ action reporting is located primarily within the West or accountable to Western standards of publication or funding (or both). Owing to linguistic and other administrative barriers, local people who deal with what the West calls disaster do not often have access to these materials. Beninese philosopher Paulin Hountondji (2009, p. 128) summarises similar concerns in the context of African studies:

> our scientific activity is extraverted, i.e. externally oriented, intended to meet the theoretical needs of our Western counterparts and answer the questions they pose. The exclusive use of European languages as a means of scientific expression reinforces this alienation. The majority of our country people are de facto excluded from any kind of discussion

about our research outcome, given that they don't even understand the languages used. The small minority who understands knows, however, that they are not the first addressees but only, if anything, occasional witnesses of a scientific discourse meant primarily for others. To put it bluntly, each African scholar has been participating so far in a vertical discussion with his/her counterparts from the North rather than developing horizontal discussions with other African scholars.

As a result, in the scholarship on disasters as well as in African studies, there is no real opportunity for those deemed at risk to raise any concerns about the veracity of the Western discourses on disaster. Lyotard (1979, p. 77) expands on how power legitimates knowledge and truth:

> la "réalité" étant ce qui fournit les preuves pour l'argumentation scientifique et les résultats pour les prescriptions et les promesses d'ordre juridique, éthique et politique, on se rend maître des unes et des autres en se rendant maître de la "réalité", ce que permettent les techniques. En renforçant celles-ci, on "renforce" la réalité, donc les chances d'être juste et d'avoir raison. Et, réciproquement, on renforce d'autant mieux les techniques que l'on peut disposer du savoir scientifique et de l'autorité décisionnelle. Ainsi prend forme la légitimation par la puissance[16].

Such skewed power relations are evident in what Virginia Garcia-Acosta (2009) referred to as *tortícolis académica*[17] or in what Gyan Prakash (1994, p. 1484) called asymmetric ignorance: 'non-Westerners must read "great" Western historians (...) to produce the good histories, while the Western scholars are not expected to know non-Western work'. Another great Indian historian, Dipesh Chakrabarty (2000, p. 28), adds:

> "they" [Western historians] produce their work in relative ignorance of non-Western histories, and this does not seem to affect the quality of their work. This is a gesture, however, that "we" [Indian historians] cannot return. We cannot even afford an equality or symmetry of ignorance at this level without taking the risk of appearing "old-fashioned" or "outdated".

This holds true in disaster studies too. This skewed representation of disaster outside of the West is self-sustaining. Indeed, the growing body of scholarship driven by Western ontologies and epistemologies as well as increasing 'humanitarian' encounters between Western and non-Western practitioners reinforces the divide between the 'safe' West and the 'dangerous' rest of the world. Western scholars and practitioners expect to encounter what they mean by disaster and indeed encounter such disasters so that in the end they strengthen the dominant discourses and provide a basis for more research and more 'humanitarian interventions' to foster disaster risk reduction in places that allegedly lack the appropriate knowledge and resources.

In both relationships, researcher versus researched and Western versus non-Western researchers, re-presentation and representation combine to create hegemonic discourses of disasters centred on Western ontologies and epistemologies. In the words of Mudimbe (1988, p. 15),

> in the name of both scientific power and knowledge, it reveals in a marvellous way (...) an epistemological ethnocentrism; namely, the belief that scientifically there is nothing to be learned from "them" unless it is already "ours" or comes from "us".

Such hegemony excludes non-Western and subaltern voices, including those of the people affected by what we call disasters, as discussed in Chapter 7, and raises axiological questions, especially with regard to the ethics of disaster research, in Deleuze's terms '*l'indignité de parler pour les autres*'[18] (Foucault and Deleuze, 1972, p. 5). As Wittig (1980, p. 106) summarises:

> There is nothing abstract about the power that sciences and theories have, to act materially and actually upon our bodies and our minds, even if the discourse that produces it is abstract. (...) All of the oppressed know this power and have had to deal with it. It is the one which says: you do not have the right to speech because your discourse is not scientific and not theoretical, you are on the wrong level of analysis, you are confusing discourse and reality, your discourse is naive, you misunderstand this or that science.

Towards a postcolonial disaster studies agenda

Exploring what and who is left out of the dominant discourses in disaster studies, and their extension to disaster risk reduction, is a task that one may associate with the wide spectrum of postmodern, poststructuralist and postcolonial theories. It is not our intention to provide an overview or summary of such diverse and contested fields of scholarship. However, a number of shared threads of thinking provide useful pathways to support our argument.

The first common trait of these branches of critical theory relevant to our approach is that they all challenge the alleged centrality of Eurocentric ontologies and epistemologies in understanding the world, including disasters, or in Young's (2004, p. 51) words that 'European culture's awareness that it is no longer the unquestioned and dominant centre of the world'. In short, postmodern, poststructuralist and postcolonial theories mark the end of the hegemonic *métarécits* inherited from the Enlightenment (Lyotard, 1979).

The second common attribute of postmodern, poststructuralist and postcolonial theories relevant here is that they all contest representations of the world through the lens of binaries, such as those inherited from structuralist theory (Bhabha, 1994; Mbembe, 1992). Not only do they allow us to go beyond the nature/hazard–culture/vulnerability binary that underpins disaster studies, they further provide the theoretical space to challenge a myriad of other dichotomies inspired by Western interpretations of the world that have emerged upon and around the nature–culture binary (Castree, 2014), including to justify and sustain normative disaster risk reduction policies and actions. The following chapters of this book will confront the relevance of the West/North–East/South and centre–periphery/margin dialectics (Chapters 3, 5 and 7), the tensions between insiders and outsiders (Chapter 3 and 7), the rift between bottom-up and top-down strategies (Chapters 5, 6 and 7), the divide between local and scientific knowledge (Chapter 7) and finally the male/man–female/woman binary (Chapter 8).

Our approach further focuses on in-betweens (Bhabha, 1994), interstices (Prakash, 1994) and frontiers (Foucault, 1969b, 1984a) and on those voices, processes and things that fall through the cracks of the *métarécits* (Spivak, 1993a). In this perspective, we prefer framing our agenda within a postcolonial rather than a decolonising perspective. The latter entails a dialectical relationship between the colonised and the coloniser (Constantino, 1978; Mignolo, 2011; Tuhiwai-Smith, 2012; Verges, 2019). Postcolonial theories rather encourage 'a movement beyond a relatively binaristic, fixed and stable mapping of power relations between "colonizer/colonized" and "centre/periphery". Such rearticulations suggest a more nuanced discourse which allows for movement, mobility and fluidity' (Shohat, 1992, p. 108) and provides space to capture the very diverse realities of the world.

Focusing on in-betweens, interstices and frontiers also allows us to reveal *contre conduites* and strategies of resistance in all their diversity, including within the West and institutions that epitomise its power, such as international organisations. These reflect how people maximise available resources and skills in a way that transcends binary views of the world (Bhabha, 1994), which we tackle in our approach to gender beyond the women–men binary in Chapter 8. In some other instances, these strategies of resistance allow local people to subvert or reinterpret techniques of control and turn them to their benefit (Scott, 1985, 1990), as in the case of people imprisoned in the Philippines, which we discuss in Chapter 9. It is our intention to study these strategies of resistance, not only in their very local relevance but also within the broader context of unequal power relations and related interstices that favour their emergence. As Shohat (1992, p. 109) recommended, 'a celebration of syncretism and hybridity per se, if not articulated in conjunction with questions of hegemony and neo-colonial power relations, runs the risk of appearing to sanctify the fait accompli of colonial violence'.

Identifying interstices and silences therefore requires us to disassemble the hegemonic Western discourses on what we call disaster. Such an endeavour of deconstruction is essential to unpack the nexus between power and knowledge in both disaster studies and disaster risk reduction. It ultimately allows us to challenge what is taken for granted in order to dissociate, on the one hand, the concept from the word disaster and, on the other hand, both the concept and the word from what is the actual reality on the ground. According to Derrida (1983, pp. 40–41),

> every conceptual breakthrough amounts to transforming, that is to say deforming, an accredited, authorized relationship between a word and a concept, between a trope and what one had every interest to consider to be an unshiftable primary sense, a proper, literal or current usage.

This process or 'event' of deconstruction – to stick to Derrida's (2003) own words – is not an act of demolition and destroying. It is a process of disassembling to better understand how things work, which we endeavour to apply to critiquing disaster studies and disaster risk reduction here. As such, we consciously diverge from Derrida's (1967a) initial agenda to deal with texts and logos. Rather, we locate our approach within the broader applications of deconstruction that have emerged among Anglo-American scholars, including in postcolonial studies (Spivak, 1987). We converge with Young's (2004, p. 51) agenda to provide 'a deconstruction of the concept, the authority, and assumed primacy of, the category of the West'. This is an endeavour that Derrida (2004b) himself endorsed in his last interview: '*la déconstruction en général est une entreprise que beaucoup ont considérée, à juste titre, comme un geste de méfiance à l'égard de tout eurocentrisme*'[19].

Deconstruction has to happen from within the system – that is, from within the hegemonic discourses that make it possible for Western understandings of disaster to sustain a normative approach to disaster risk reduction. This endeavour is therefore a matter of investigating from within the articulation of Western ontologies and epistemologies and their impact on policies and action to reduce disaster risk. Derrida (1967, p. 39) justifies this process as follows:

> Les mouvements de déconstruction ne sollicitent pas les structures du dehors. Ils ne sont possibles et efficaces, ils n'ajustent leurs coups qu'en habitant ces structures. En les habitant d'*une certaine manière*, car on habite toujours et *plus encore quand on ne s'en doute pas*. Opérant nécessairement de l'intérieur, empruntant à la structure ancienne toutes les ressources stratégiques et économiques de la subversion, les lui empruntant structurellement, c'est-à-dire sans pouvoir en isoler des éléments et des atomes, l'entreprise de déconstruction est toujours d'*une certaine manière* emportée par son propre travail[20].

As such, the concept of disaster is an overarching node that constitutes an essential reference for our approach. Our intention is to critique its scope, interpretation and application as well as those of cognate concepts such as hazard, vulnerability, capacities and resilience in order to highlight ambiguities and incoherence in its use in various contexts. As Spivak (1987, p. 103) said: 'no rigorous definition of anything is possible but definitions are necessary to keep us going and take a stand'. We therefore follow Butler (1995, p. 51), who once suggested that 'to deconstruct these terms means, rather, to continue to use them, to repeat them, to repeat them subversively, and to displace them from the contexts in which they have been deployed as instruments of oppressive power'. By doing so, we hope to uphold Derrida's initial call (1967, p. 25):

> A l'intérieur de la clôture, par un mouvement oblique et toujours périlleux, risquant sans cesse de retomber en-deçà de ce qu'il déconstruit, il faut entourer les concepts critiques d'un discours prudent et minutieux, marquer les conditions, le milieu et les limites de leur efficacité, désigner rigoureusement leur appartenance à la machine qu'ils permettent de constituer; et du même coup la faille par laquelle se laisse entrevoir, encore innommable, la lueur de l'outre-clôture[21].

Such an agenda is nothing new in many of the social sciences, literary studies and philosophy. Nonetheless, it has never really trickled through the otherwise porous borders of disaster studies, and there are very few existing attempts to consider (i.e. deconstruct) disaster, disaster studies and disaster risk reduction from a postcolonial perspective (e.g. Carrigan, 2010, 2015; Alburo-Cañete, 2021; Lakhina, 2020; Rastogi, 2020) and not many more that embrace, either fully or partially, a broader poststructuralist perspective (e.g. Protevi, 2006, 2009; Clark, 2011; Marchezini, 2015; Barrios, 2017; Covarrubias and Raju, 2020). This is the approach we are taking in this book.

Our book and its approach

In writing this book and venturing down this postcolonial path, we hope to uphold calls made by some of our mentors back in the 1970s and 1980s who then aimed to revisit disaster studies and challenge assumptions about what a disaster is (Wisner et al., 1976, 1977; Waddell, 1977; Lewis, 1976a, 1979; Hewitt, 1983). We want to emphasise the importance of looking back in time and engaging with seminal, but sometimes forgotten, works. We also aim to affirm the importance of engaging with theory, especially critical theory from the West and elsewhere, to inform research on disaster, but also theory that helps in informing disaster risk reduction in a meaningful perspective. As Deleuze once said in his famous dialogue with Foucault: theory '*il faut que ça serve, il faut que ça fonctionne*'[22] (Foucault and Deleuze, 1972, p. 5).

As such, we aim both to critically discuss and deconstruct the dominant ontologies and epistemologies that sustain the hegemony of Western discourses in disaster studies and *dispositif* of disaster risk reduction and to reveal how these discourses and *dispositif* fail to recognise the reality of everyday lives, personal priorities and intrinsic abilities to deal with what we call 'disaster'. We will henceforth use the term disaster in a critical perspective, not as an antithetical reproduction of a Western concept but as the aim of our enterprise of deconstruction. It is, again, only through its own lens that its scope, relevance and flaws can be revealed. As Butler (1995, p. 49) suggested, 'to deconstruct is not to negate or to dismiss, but to call into question and, perhaps most importantly, to open up a term (...) to a reusage or redeployment that previously has not been authorized'.

Our effort is therefore located mainly within the realm of the dominant Western ontologies and epistemologies that dominate disaster studies, in order to question their intrinsic nature, scope and relevance. We thus explicitly position our critique within the space of the nature/hazard–culture/vulnerability binary that we exposed earlier in this introductory chapter. This is why we borrow from Western scholarship to support our argument and sustain our deconstruction agenda. Such an approach is prone to criticism, as acknowledged by Derrida (1967a) and Spivak (1993a). However, it is part of the process of disassembling from within that makes it epistemologically coherent to challenge the relevance of Western ontologies and epistemologies through resorting to this very tradition of scholarship. In addition, it is our contention later in this book that there is no such thing as a single non-Western understanding of disaster, so that the only relevant approach to stay away from any dialectical aporia is to underline the organic and intrinsic flaws of the hegemonic discourses in disaster studies and disaster risk reduction from within.

At this point, one may argue that considering Derrida and Foucault within the same framework is, if not antithetical, at least controversial as both thinkers have been assumed to express more differences than convergences in their methods and arguments (e.g. Said, 1978). However, Spivak (1993a) and Boyne (1990) have shown that Derrida and Foucault, though definitely different, can actually be read together, to use Spivak's wording, especially when it comes to analysing the nexus between knowledge and power that is at the centre of our interest. Spivak (1993a, p. 37) explains:

> Pouvoir/savoir, then, is catachrestic in the way that all names of processes not anchored in the intending subject must be: lines of knowing constituting ways of doing and not doing, the lines themselves irregular clinamens from subindividual atomic systems – fields of force, archives of utterance. Inducing them is that moving field of shredded énoncés or differential forces that cannot be constructed as objects of investigation. Ahead of them, making their rationality fully visible, are the great apparatuses of puissance/connaissance. Between the first and the second there is the misfit of the general and the narrow sense. Between the last two is the misfit that describes examples that seem not to be faithful to the theorist's argument. If read by way of the deconstructive theorizing of practice, this does not summon up excuses or accusation. This is how theory brings practice to crisis, and practice norms theory, and deviations constitute a forever precarious norm; everything opened and menaced by the risk of paleonymy. Thus I give the name of Foucault in to Derrida.

Our endeavour of deconstruction from within also allows us to focus on the superstructure which overarches the dominant discourses and *dispositif*, in our case on an international level, and the underpinning infrastructure, both of which are the actual modes of production of knowledge in disaster studies and the unequal power relations among the stakeholders of disaster risk reduction. As such, our endeavour of deconstruction addresses one of Althusser's (2018) critiques addressed to Gramsci's approach to hegemony. We believe that focusing on the infrastructure, including the actual modes of production of knowledge, the unequal power relations they mirror and the *dispositif* of disaster risk reduction they underpin, also contributes to addressing some of the most radical, Marxist critiques of postcolonial theory, such as those mounted by Vivek Chibber (2013), Ahmad (1992), Dirlik (1997), San Juan (1998) and Amselle (2008).

In fact, we position our postcolonial approach of disaster beyond the sole study of texts which were both the primary interest of pioneer postcolonial scholars and the target of their Marxist critiques. We take our agenda to the study of discourses on nature and culture/society for their importance, as we saw earlier in this chapter, in underpinning disaster studies and the governmentality of disaster. We follow Foucault in considering such discourses beyond the sole remit of texts, and even speeches, to actually capture the broader *dispositif* of disaster risk reduction as shaped by Western scientific paradigms. Our intention is to uncover why the hegemony of Western knowledge and the governmentality of disaster fail to capture the diverse realities of the world, indeed globalised – but not homogenised – to the point that all cultural and social differences have been erased.

It is in this perspective that the following eight chapters successively explore ways of knowing in disaster studies, strategies of disaster risk reduction, and strategies of resistance within the scope of familiar concepts, methods and scientific paradigms. It is only in the final essay (Chapter 10) that we step back from this ontological assumption to initiate a critique from the outside and expand our opening questioning as to whether there is actually such a thing as a disaster. It is our contention that discourses on disaster are inherently a Western invention as they are mediated by concepts, methods, scientific narratives, norms and values that are those of the West. They contribute to depict, capture and make tangible the very diverse experiences of millions of people across locations and cultures under the supposed universality of the nature/hazard–culture/binary.

In doing so, we recognise that the notion or label of the 'West' is problematic and contested in many ways. It is, in itself, a construct that is part of a binary opposition with the rest of the world that early postcolonial thinkers such as Fanon, Memmi and Said used to support their own analysis as much as to devise strategies of resistance and decolonisation (Young, 1995b). We endeavour here to follow the path of Spivak (1987) and Bhabha (1994), who consider the West beyond this dialectic and within a multi-polar complex of interactions that allows one particular set of ontologies and epistemologies to rule disaster studies and disaster risk reduction policy and actions.

Similarly, we do not assume that there is such a thing as a single homogenous West (Young, 1995b; Chakrabarty, 2000), nor is there a single Western understanding of disaster. The very existence of different paradigms and discourses clearly shows that. However, as discussed above, all of these paradigms and discourses are polarised around the nature–culture binary that reflects an *épistémè* inherited from the Enlightenment. It is in this ontological and epistemological legacy that we track a common Eurocentric heritage that we call the West[23]. As Edouard Glissant (1981, p. 12) aptly stated, 'l'Occident n'est pas à l'ouest. Ce n'est pas un lieu, c'est un projet'.

Furthermore, our intention is not to dismiss the immense contribution of the Enlightenment to Western science or to the understanding and enhancing of the contemporary everyday lives of people in Europe. Nor is our aim to debunk the relevance of Western understandings of disaster within the West, especially in Europe. We endeavour to go beyond such a trivial approach and take on Foucault's (1984a) recommendation:

> il faut refuser tout ce qui se présenterait sous la forme d'une alternative simpliste et autoritaire: ou vous acceptez l'Aufklärung, et vous restez dans la tradition de son rationalisme (ce qui est par certains considéré comme positif et par d'autres au contraire comme un reproche); ou vous critiquez l'Aufklärung et vous tentez alors d'échapper à ces principes de rationalité (ce qui peut être encore une fois pris en bonne ou en mauvaise part). Et ce n'est pas sortir de ce chantage que d'y introduire des nuances "dialectiques" en cherchant à déterminer ce qu'il a pu y avoir de bon et de mauvais dans l'Aufklärung[24].

Our point is to challenge the undiscerned and imperialistic rolling-out of Western discourses on disasters into contexts where they do not make much sense. We further endeavour to unpack how ethnocentrism in disaster studies has shaped our understanding of disasters in a way that has excluded many diverse non-Western epistemologies and ontologies and, as a result, failed to recognise the realities of most people affected by what we call disasters. In other words, we set out to understand why it is that Western ontologies and epistemologies have become the only apparent resources to study disasters and design disaster risk reduction policies and actions independently of their geographical and cultural context. As such,

> the point is not that Enlightenment rationalism is always unreasonable in itself but rather a matter of documenting how – through what historical process – its "reason", which was not always self-evident to everyone, has been made to look "obvious" far beyond the ground where it originated.
>
> (Chakrabarty, 2000, p. 20)

Finally, this book is asking more questions than it provides definite answers. It deconstructs dominant discourses but does not propose a definitive alternative approach to subvert it. The main reason lies in the intrinsic nature of deconstruction: it is an enterprise of disassembling and questioning (Derrida, 1967a). Such endeavour nonetheless constitutes a first, required step so that the incoherence and flaws of Western discourses in disaster studies and their influence on disaster risk reduction can be exposed. If our final essay suggests some leads and perspectives for overcoming the incoherence and flaws, it is, however, not the task for a single book, nor is it for a single initiative to find solutions for the very reason that such a proposal would be antithetical with the pluralistic nature of our postcolonial agenda. Alternatives will have to emerge from below, within specific contexts to reflect local ontologies and epistemologies.

Notes

1 *'Since these concepts are indispensable for unsettling the heritage to which they belong, we should be even less prone to renounce them'*. From the English version translated by Gayatri Chakravorty Spivak and published by the Johns Hopkins University Press in 1974.

2 *'It is thinking about the origin and scope of the question "what is?…", the question that governs the whole history of philosophy. Every time we try to think of the possibility of the "what is?…", to ask a question on this form of question, or to question the necessity of this wording in a specific language, a certain tradition, etc., what we do at this point only addresses the question "what is?" to a certain extent. This is the difference of deconstruction. It is an interrogation on everything that is more than an interrogation'*. Our translation.

3 *'Power is everywhere; not because it embraces everything, but because it comes from everywhere. (…) Power is not an institution, and not a structure; neither is it a certain strength we are endowed with; it is the name that one attributes to a complex strategical situation in a particular society'*. From the English version translated by Robert Hurley and published by Pantheon Books in 1978.

4 *'A thoroughly heterogeneous ensemble consisting of discourses, institutions, architectural forms, regulatory decisions, laws, administrative measures, scientific statements, philosophical, moral and philanthropic propositions – in short, the said as much as the unsaid. Such are the elements of the apparatus. The apparatus itself is the system of relations that can be established between these elements. (…) a sort of – shall we say – formation which has as its major function at a given historical moment that of responding to an urgent need. The apparatus thus has a dominant strategic function'*. From the English version translated by Colin Gordon and published by Pantheon Books in 1980.

5 *'Enlightenment is man's emergence from his self-imposed nonage. Nonage is the inability to use one's own understanding without another's guidance. This nonage is self-imposed if its cause lies not in lack of understanding but in indecision and lack of courage to use one's own mind without another's guidance. Dare to know! Have the courage to use your own understanding, is therefore the motto of the enlightenment'*. From the English version translated by Mary C. Smith and published by Columbia University: http://www.columbia.edu/acis/ets/CCREAD/etscc/kant.html#note1.

6 *'He must not blame providence in any way for the troubles that harm him (…) but that he ought to recognize every single event as if it in all respects were produced by himself, and that he therefore must accept himself the full responsibility for his own hardship'*. From the English version translated by Robert B. Louden and published by Cambridge University Press in 2007.

7 *'the largest of all by its commerce, by its navigation, by its fertility, by the Enlightenment & the industry of its people, by the knowledge of arts, sciences and 'trades'*. Our translation.

8 '*The awakened and troubled consciousness of modern thought*'. From the English version published by Pantheon Books in 1970.

9 '*The function of this center was not only to orient, balance, and organize the structure – one cannot in fact conceive of an unorganized structure – but above all to make sure that the organizing principle of the structure would limit what we might call the play of the structure*'. From the English version translated by Alan Bass published by the University of Chicago Press in 1978.

10 '*The colonialists are perpetually explaining, justifying and maintaining (by word as well as by deed) the place and fate of their silent partners in the colonial drama. The colonized are thus trapped by the colonial system and the colonialist maintains his prominent role. Colonial racism is built from three major ideological components: one, the gulf between the culture of the colonialist and the colonized; two, the exploitation of these differences for the benefit of the colonialist; three, the use of these supposed differences as standards of absolute fact*'. From the English version translated by Howard Greenfeld published by Earthscan in 1974.

11 '*Since the 19th century, we have to say that ways of thinking, political schemes, fundamental economic mechanisms that were those of the West have become universal, through the violence of colonisation, or let's say, most of the time, have, as a result, become universal. And this is why I mean by the West, this sort of small portion of the world which strange and violent destiny has been to ultimately impose its ways of seeing, thinking, saying and doing to the whole world*'. Our translation.

12 '*The calcified colonized society is therefore the consequence of two processes having opposite symptoms: encystment originating internally and a corset imposed from outside*'. From the English version translated by Howard Greenfeld published by Earthscan in 1974.

13 '*The white man takes his own mythology (that is, Indo-European mythology), his logos-that is, the mythos of his idiom, for the universal form of that which it is still his inescapable desire to call Reason*'. From the English version translated by F.C.T. Moore and published in New Literary Theory in 1974.

14 '*This discourse of platitude has led to false oppositions in the name of which (Western and indigenous) commentators feel obliged to resort to either clichés on alleged cultural specificities (forgetting that they may only be a language on the very prosaic modes of social control) or to a judgment of the "Other" in light of so-called Western values that we project on the object of research, thus omitting to capture actual equilibriums that result from different cultures and histories*'. Our translation.

15 '*They cannot represent themselves, they must be represented. Their representative must at the same time appear as their master, as an authority over them, an unlimited governmental power which protects them from the other classes and sends them rain and sunshine from above*'. From the English version translated by Saul K. Padover and published by Progress Publishers in 1937.

16 '"*Reality" is what provides the evidence used as proof in scientific argumentation, and also provides prescriptions and promises of a juridical, ethical, and political nature with results, one can master all of these games by mastering "reality". That is precisely what technology can do. By reinforcing technology, one "reinforces" reality, and one's chances of being just and right increase accordingly. Reciprocally, technology is reinforced all the more effectively if one has access to scientific knowledge and decision-making authority. This is how legitimation by power takes shape*'. From the English edition translated by Geoff Bennington and Brian Massumi and published by University of Minnesota Press in 1984.

17 *academic torticollis*

18 '*The indignity of speaking for others*'. From the English version published in Telos in 1973.

19 '*Deconstruction in general is an undertaking that many have considered, and rightly so, to be a gesture of suspicion with regard to all Eurocentrism*'. From the English version translated by Pascale-Anne Brault and Michael Naas and published by Palgrave Macmillan in 2007.

20 '*The movements of deconstruction do not destroy structures from the outside. They are not possible and effective, nor can they take accurate aim, except by inhabiting those structures. Inhabiting them in a certain way, because one always inhabits, and all the more when one does not suspect it. Operating necessarily from the inside, borrowing all the strategic and economic resources of subversion from the old structure, borrowing them structurally, that is to say without being able to isolate their elements and atoms, the enterprise of deconstruction always in a certain way falls prey to its own work*'. From the English version translated by Gayatri Chakravorty Spivak and published by the Johns Hopkins University Press in 1974.

21 '*Within the closure, by an oblique and always perilous movement, constantly risking falling back within what is being deconstructed, it is necessary to surround the critical concepts with a careful and thorough discourse-to mark the conditions, the medium, and the limits of their effectiveness and to designate rigorously their intimate relationship to the machine whose deconstruction they permit; and, in the same process, designate the crevice through which the yet unnameable glimmer beyond the closure can be glimpsed*'. From the English version translated by Gayatri Chakravorty Spivak and published by the Johns Hopkins University Press in 1974.

22 '*It must be useful. It must function*'. From the English version published in Telos in 1973.

23 '*The West is not in the West. It is a project, not a place*'. From the English version translated by J. Michael Dash and published by University Press of Virginia in 1989.

24 '*One has to refuse everything that might present itself in the form of a simplistic and authoritarian alternative: you either accept the Enlightenment and remain within the tradition of its rationalism (this is considered a positive term by some and used by others, on the contrary, as a reproach); or else you criticize the Enlightenment and then try to escape from its principles of rationality (which may be seen once again as good or bad). And we do not break free of this blackmail by introducing "dialectical" nuances while seeking to determine what good and bad elements there may have been in the Enlightenment*'. From the English version translated by Catherine Porter and published by Pantheon Books in 1984.

2 A genealogy of disaster studies

This chapter explores the origin and theoretical foundations of the current discourses on disasters. It does not intend to track the origin of disaster studies as a linear heritage. Rather, it takes a genealogical approach in Foucault's (1971) sense; that it is to reveal the ontological and epistemological conditions for the emergence of concepts and the mode of formation, sequence and co-existence of associated discourses. As such, we not only track the origins of modern disaster studies to the 1755 Lisbon earthquake but also contextualise its emergence within the particular perspective of the Enlightenment and its legacy gathered in what Foucault (1966) called the modern *épistémè* of the West. It is this heritage that transpires in the ways we understand and study disasters almost three centuries later. It is also this heritage and the alleged prominence that the West has given to its science that underpins the contemporary hegemony of Eurocentric/Western discourses in both disaster studies and disaster risk reduction. In the subsequent sections, we unpack the origin of the ontological assumption that disasters sit at the interface between nature and culture as celebrated by the mnemonic of 'disaster = hazard × vulnerability'. We extend this discussion to cognate concepts such as capacities and resilience. These concepts eventually provide an entry point to study the conditions of the emergence of scientific paradigms and how these paradigms underpin broader discourses on disasters.

On the origins of modern disaster studies

Disaster is obviously a term of Western origin, more precisely of Latin etymology combining the prefix *Dis* and the root word *Astrum*, which literally means the 'unfavourable aspect of a star'. The *Oxford Dictionary of English Etymology* indicates that it travelled into English around the sixteenth century through the Italian word *disastro* and its French equivalent *désastre*. However, neither the *Encyclopédie* of Diderot and d'Alembert nor Chambers's *Cyclopædia* had a specific entry for the term in the early eighteenth century. Nor did they have entries for the cognate term 'calamity', the origin of which similarly stems from the Latin *calamitās*. Both encyclopaedias do have entries for 'catastrophe', from the Latin *catastropha* and the Greek *katastrophé*, but they refer to the brutal twist at the end of a poem or piece of drama.

DOI: 10.4324/9781315752167-2

Although Defoe's (1704) landmark report on the storm that hit England and Wales in 1703 signalled an early popularisation of the term disaster, things really changed with the 1755 Lisbon earthquake. The entries for both 'Lisbon' and 'earthquake' in the *Encyclopédie* were actually written after the 1755 earthquake and there the term *désastre* is clearly associated with the event, marking a turning point in our understanding of what we now call disaster. This turning point has been most famously captured in Rousseau's (1967, p. 39) critique of Voltaire's *Poème sur le désastre de Lisbonne*.

> (…) convenez, par exemple, que La nature n'avait point rassemblé Là vingt mille maisons de six à sept Etages, et que si Les habitans de cette grande ville eussent été dispersés plus également et plus légèrement logés, le dégat eut été beaucoup moindre, et peut-être nul. Tout eut fui au premier ébranlement, et on les eut vus le lendemain à vingt lieües de là, tout aussi gais que s'il n'était rien arrivé. Mais il faut rester, s'opiniâtrer autour des masures, s'exposer à de nouvelles sécousses, parce que ce qu'on laisse vaut mieux que ce qu'on peut emporter. Combien de malheureux ont péri dans ce désastre pour vouloir prendre, L'un ses habits, l'autre ses papiers, l'autre son argent?[1]

Rousseau not only introduces a human/social dimension to our understanding of disasters and their impact but also moves the analysis beyond the representation of natural hazards and their impact – such as in Defoe's (1704) report on the storm of 1703 – to an interrogation of causes as a historical process. This shift in thinking, Foucault (1966) would argue, initiated Europe's transition from the classic to the modern *épistémè*.

In his *Essai sur l'origine des langues*, Rousseau (1969, p. 113) would eventually establish a dialectical and historical relationship between nature and culture, which sets disasters outside of the everyday of both nature and society:

> Les associations d'hommes sont en grande partie l'ouvrage des accidens de la nature: les déluges particuliers, les mers extravasées, les éruptions des volcans, les grands tremblemens de terre, les incendies allumés par la foudre et qui détruisaient les forêts, tout ce qui dut ensuite effrayer et disperser les sauvages habitans d'un pays, dut ensuite les rassembler pour réparer en commun les pertes communes: les traditions des malheurs de la terre, si fréquens dans les anciens temps, montrent de quels instrumens se servit la Providence pour forcer les humains à se rapprocher. Depuis que les sociétés sont établies, ces grands accidens ont cessé et sont devenus plus rares: il semble que cela doit encore être; les mêmes malheurs qui rassemblèrent les hommes épars disperseraient ceux qui sont réunis.[2]

By clearly positioning disasters at the interface between nature and culture and by establishing a dialectical relationship between the two, Rousseau pre-empted contemporary analyses of disasters that would eventually be structured around the widely accepted binary mnemonic of 'disaster = hazard × vulnerability' (and all its cognate versions). Throughout the nineteenth and twentieth centuries, this polarised perspective on disaster and the continuing interest in uncovering their causes would firmly ground disaster studies in the modern *épistémè* of the West, which has centred on simultaneously controlling Nature and rationalising society to set humankind free and allow it to flourish.

The Lisbon earthquake occurred during an intense period of seismicity in Europe and a concurrent quest for better understanding nature in its multiple facets. The eighteenth century, or *Siècle des Lumières*, saw the strengthening and systematic study of the natural environment through an increasingly fragmented field of diverse sciences that would each claim their own space and object of study. The publication of encyclopaedias and seminal treaties would systematise and rationalise this emerging knowledge of the world around us (Foucault, 1966), including that of natural hazards as per the works of Kant, von Humboldt and many others. In this context, natural hazards, especially the most extreme and spectacular such as earthquakes and volcanic eruptions, provided a powerful springboard for justifying and furthering scientific endeavours, as it was only through better knowledge that humankind was to be freed of its fears and allowed to enter an era of liberation and flourishment: '*die Verdoppelung der Natur in Schein und Wesen, Wirkung und Kraft, die den Mythos sowohl wie die Wissenschaft erst möglich macht, stammt aus der Angst des Menschen, deren Ausdruck zur Erklärung wird*'[3] (Horkheimer and Adorno, 1947, p. 21).

Controlling nature and allowing people to live away from harm were therefore at the centre of the Enlightenment project of liberation of humankind, the '*absoluten Lebenszweck*'[4] (Horkheimer and Adorno, 1947, p. 38). The Lisbon earthquake emphasised that enhanced understanding of hazards would allow for a rational planning of cities and the application of reason to the more general organisation of people's lives and livelihoods. This rationalisation of lives and living environments through a new form of governmentality would facilitate the later emergence of disaster policies polarised between nature and culture. Policies that would centre on progress and modernity.

The divide between nature and culture and polarised views of disaster across the hazard–vulnerability binary are essential to understanding how disaster studies have emerged, developed and strengthened over the past century. Dominant concepts and scientific paradigms as well as policies and actions towards disaster risk reduction will all be located within this ontological space and be informed by epistemologies inherited from the Enlightenment.

A myriad of cognate concepts

Disaster studies have been informed by a number of concepts, all of them located within the dialectical relationship established between nature and culture, and formalised in the mnemonic 'disaster = hazard × vulnerability'.

With the expansion of scholarship on the social side of the binary in the second half of the twentieth century, cognate concepts have eventually been added to capture the multifaceted nature of people's lives and livelihoods. These additional concepts most notably include capacities and resilience. It is not our intention to provide an exhaustive review of these concepts, nor do we aim to unpack the diversity of definitions and framings they reflect. There are far too many. Neither do we focus on risk as we consider that it does not mark any major theoretical breakthrough since it often stands for (or in combination with) disaster before losses occur. Rather, our intention is to provide sufficient overview to emphasise the dialectical relationship between nature and society as the foundation for the different scientific paradigms that have supported broader discourses on disasters and disaster risk reduction.

The concept of hazard has been around, in its contemporary meaning, since at least the sixteenth or seventeenth century. It is assumed to have come to the English language from the Arabic *az-zhar* (gaming dice) through the Spanish *azar* and the French *hasard*. As such, it is not a Western concept per se but its contemporary meaning relevant to disaster studies definitely is. Both Chambers's *Cyclopædia* and Diderot and D'Alembert's *Encyclopédie* actually refer to hazard as a threat to people and their livelihoods. The *Encyclopédie* is a bit more specific, pointing to hazard as '*des évenemens, pour marquer qu'ils arrivent sans une cause nécessaire ou prévûe*'.[5] If these losses were associated primarily with the hazard of maritime trade and travel, they were also linked to natural phenomena, which receive significant attention in both encyclopaedias, which have specific and detailed entries for volcanic eruptions, *soufrières*, earthquakes, floods, hurricanes, storms and deluges. These days, the word hazard similarly refers to the phenomenon of potential threat to people and their livelihoods in a particular place and at a particular point in time (Wisner et al., 2012). In the context of this book, the word refers to those associated with the natural environment in particular. The occurrence of these hazards has long been attributed solely to nature and its extremes. For example, Burton and Kates (1964, p. 413) underline 'those elements in the physical environment, harmful to man and caused by forces extraneous to him'. Frampton et al. (2000, p. 3) further stress an 'uncontrollable' dimension of the physical event, whereas Chapman (1994, p. 1) refers to 'an extreme or rare natural phenomenon (...) greatly exceeding normal human expectation in terms of its magnitude or frequency'. The view that hazards are extraneous to society has been revisited to consider an anthropogenic contribution to flooding, landslides and other phenomena. However, most definitions continue to associate hazards with extreme and rare natural phenomena which exceed the ability of people to remain unscathed.

The concept of vulnerability stems from the Latin *vulnerãbilis* (wounding); it travelled into English with its contemporary meaning around the seventeenth century and eventually arrived as a standalone entry of Diderot and D'Alembert's *Encyclopédie* to qualify what can be wounded. The concept was introduced in the disaster literature in the 1970s (e.g. Baird et al., 1975; O'Keefe et al., 1976; Wisner et al., 1977) with similar definitions that revolved around the susceptibility to suffer damage should a hazard occur. Vulnerability was then a social construct, stressing the conditions of a society that makes it possible for a hazard to become a disaster (Cannon, 1994, p. 13). The concept was pitched to allow the identification of hazard-independent processes that make some people more fragile than others when hazards strike (Lewis, 1999; Watts and Bohle, 1993; Wisner et al., 2004). Originally, vulnerability was thus underpinning constructivist and qualitative enquiries. However, other definitions later focused on 'community' or territorial scales to draw vulnerability maps through the integration of quantitative and qualitative data dealing with the actual condition of the insecurity of people in the face of natural hazards (e.g. Anderson-Berry, 2003; Birkmann, 2006). This approach emerged with the desire for measuring vulnerability and making it quantitative for disaster risk reduction. Engineers and earth scientists have also used, since at least the late 1970s, the concept of vulnerability to compute quantitative indices of potential losses of built structures should a damaging event occur (e.g. Kemp, 2007; Stewart, 2003).

The concept of capacities (plural, as itself and not associated with other concepts such as in the expression 'capacity to resist', 'to face' and 'to recover') emerged within circles of practitioners in the late 1980s (Anderson and Woodrow, 1989). It came to English from the Latin *capacitas* in the fifteenth century through the French *capacité* and was captured in both the *Cyclopædia* and the *Encyclopédie* with the same meaning: an aptitude or disposition to do or hold something. It is in this perspective that it is used nowadays in disaster studies to reflect the increased recognition of people's ability to face natural hazards which was not captured in the mainly negative concept of vulnerability. Capacities refer to the 'set of diverse knowledge, skills and resources people can claim, access and resort to in dealing with hazards and disasters' (Gaillard et al., 2019). This knowledge and these skills and resources are individual and collective and often are shared and combined among relatives, kin and neighbours within a particular place or beyond propinquity. Since vulnerable people may or may not display a large array of capacities, capacities do not entertain a dialectical relationship with vulnerability (Davis et al., 2004). In fact, whereas the root causes of vulnerability are largely exogenous to those at risk, capacities often are rooted in resources which are endogenous to the group of people who share and combine them. They are often the extension of everyday practices and roles within the society rather than *extra*-ordinary measures taken to face rare and extreme events.

The concept of resilience similarly traces its root to the Latin verb *resilīre*, or to leap back, and was used in English with a roughly same meaning as early as the seventeenth century (Alexander, 2013). It emerged in the disaster literature in the 1970s (e.g. Torry, 1979a), then mirroring parallel use in child psychology (e.g. Werner et al., 1971), engineering (e.g. Gordon, 1978) and ecology (e.g. Holling, 1973). It spread widely in the 1990s and is still the object of a conceptual debate around its sense and application among social scientists (e.g. Alexander, 2013; Manyena, 2006). In line with Holling's ecological approach, resilience often is viewed in the long term, based on a society's evolution and ability to overcome shocks and disturbances (Folke et al., 2002). A second set of definitions and approaches looks at resilience in a narrower perspective: the flip or positive side of vulnerability or the ability to resist damage and change when facing natural hazards (Cutter et al., 2010; Turnbull et al., 2013). Pelling (2003) further considers resilience a component of vulnerability or the ability of an actor to cope with or adapt to hazards. A fourth stream of definitions place resilience in a post-disaster context to capture the ability of a system or people to recover from hazardous events (Timmermann, 1981; United Nations International Strategy for Disaster Reduction, 2002).

These subtle variations in the definitions and understandings of concepts such as resilience are ultimately of limited relevance to our argument, at least until our final essay (Chapter 10). What matters is that the concepts of vulnerability, capacities and resilience all have Latin etymologies while hazard traces its contemporary definition to a Western interpretation of an Arabic term. Furthermore, all concepts have retained a meaning roughly similar to that forged in the English language around the seventeenth century (and a bit earlier for capacities) and firmed up in the major encyclopaedias of the eighteenth century or Age of Enlightenment. As such, it is safe to posit that they all carry a strong legacy associated with Europe and its cultural heritage.

Dominant understanding and the hazard paradigm

Disasters have long been viewed through the lens of natural hazards and thus have been seen as prime concerns for physical sciences such as seismology, volcanology, climatology, geomorphology, hydrology and meteorology. Up until the mid-twentieth century, the responsibility for the occurrence of disasters therefore was attributed to external natural forces as listed in the encyclopaedias of the eighteenth century.

The 1920 PhD thesis of Samuel Prince on the Halifax disaster is widely recognised as the first piece of contemporary scholarship on the social dimension of disasters. However, it was not until the 1940s and White's (1945) pioneering thesis on the adjustment of people to floods in the United States of America, that the human dimension of disasters began to be widely accepted as part of a broader human ecology body of scholarship. Although it addresses anthropogenic issues, the human ecology approach has continued to emphasise the contribution of natural hazards and, in fact, has further consolidated the so-called hazard paradigm. This paradigm particularly stresses the importance of extreme (function of magnitude) and rare (function of time) natural hazards that exceed the ability of people to cope with them. Hazards are those often considered extraneous to people and societies.

The extraneous and extreme dimension of natural hazards leads disasters to be considered out of the regular social fabric. Scientists, institutions, governments and media often mention *extra*-ordinary, *un*-controllable, *in*-credible, *un*-predictable and *un*-certain phenomena along with *un*-expected disasters and *un*-scheduled and *un*-anticipated damage (Hewitt, 1983). Regions affected are claimed to be *un*-able to face such forces of nature and often are considered to be *under*-developed, *over*-populated, *un*-informed, *un*-prepared and *un*-planned. Therefore, a clear border is delineated between regions of the world which are often struck by disasters and those that are supposedly safe (Bankoff, 2001). The construction of this divide is facilitated by the homogenising nature of data available on the occurrence of disasters (Serje, 2012).

In this context, earth and climate scientists and engineers tend to focus on monitoring, predicting and calculating probabilities and parameters for extreme natural hazards. Hewitt (1983, p. 20) writes that 'uncertainty is the umbilical cord that ground the otherwise gratuitous notion of the accident, the separate assessment of 'extremes', in a challenging and refined language'. In parallel, social scientists, led by geographers following White's precedent, have been interested in how people and societies perceive the potential danger and how they adjust to possible threats (Burton et al., 1978; Kates, 1971). Individuals and societies said to have a low perception of risk allegedly adjust poorly to possible threats. Conversely, people and societies considered to have a high perception of risk are assumed to adjust well to natural hazards. Factors that affect people's perception of risk are hazard-related (i.e. hazard magnitude, duration, frequency and temporal spacing as well as the recentness, frequency and intensity of personal experiences with hazards) and independent from the social environment.

Proponents of the hazard paradigm further draw on behaviourist theory, inspired by Watson (1913), to stress the range of adjustment options known to those at risk. The larger the array of opportunities people are aware of, the broader their portfolio of choices to adjust to natural hazards is. Choices of adjustment therefore widen with technological progress and wealth, thus clearly aligning with the project of modernity of the Enlightenment (Kates, 1971). Poorer people are deemed to be less able to adjust to natural hazards and, as a result, suffer the brunt of disasters. In fact, in many instances, the poor are said to aggravate the impact of disasters by deliberately choosing to live in hazardous areas and contributing to higher population growth, leading to more casualties should hazards occur (Benblidia, 1990; Jones, 1981).

The study of people's perception of risk relies upon quantitative research methods, including the use of identical questionnaire surveys conducted in different regions of the world following similar approaches, so that the outcomes are comparable (White, 1974). These allow for the design of universal theoretical frameworks such as the choice tree of adjustment to natural hazard, which has been applied from individual to societal levels (Burton et al., 1978). The choice tree distinguishes incidental adjustments to natural hazards from purposeful adjustments, which are further broken down into the passive acceptance of losses, to actions to reduce losses and initiatives towards choosing – or changing – land use or location.

The hazard paradigm and study of people's perception and adjustment to natural hazards have introduced a human dimension in the study of disasters. However, because its main focus is still the hazard, it has not marked a critical shift across the nature–culture binary that underpins Western understandings of disaster. In addition, it has been widely criticized for being excessively deterministic, Malthusian and technocratic to the detriment of the study of underlying social processes that make people vulnerable.

Alternative interpretation and the vulnerability paradigm

The role of natural hazards and people's perception of associated risk in the occurrence of disasters has been progressively reconsidered since the 1970s through the work of anthropologists and geographers engaged in both political economy and political ecology research. For example, O'Keefe et al. (1976) have emphasised the root causes of people's unequal vulnerability. The bottom-line evidence is that there is no disaster which causes all buildings in a particular place to collapse or all people to die. There are always buildings that withstand and people who survive.

Drawing upon the precepts of an array of critical theories associated with the works of Marx, Webber and Sen, the proponents of the vulnerability paradigm argue that people affected by disasters are disproportionately drawn from the margins of society (Wisner et al., 2004). They are marginalised geographically and physically because they live in places exposed to natural hazards which often coincide with locations with inadequate services and contested tenure (e.g. informal settlers and shacks). Socially and culturally, they are marginalised because they are frequently members of minority groups (e.g. ethnic or caste minorities, people with disabilities, people in prison and refugees). They are economically marginalised because they are excluded from the formal job market. Ultimately, they (e.g. indigenous people, women, gender minorities, homeless, children and older people) are politically marginalised because their voice is disregarded by those with political power. Therefore, disasters most frequently affect individuals with limited and fragile incomes, reducing their ability to deal with natural hazards, notably to choose the location of their home and to invest in protective measures. Vulnerability and marginality also relate to inadequate social protection and limited social networks (Hartmann and Boyce, 1983; Wisner et al., 1976, 2012).

The unequal impact of disasters and people's vulnerability to natural hazards reflects a lack of access to resources and means of protection. Lack of access does not mean that these resources and means of protection are not available locally. In fact, they are most often available but access is limited to those with stronger economic, political and social position in society, thus reflecting an unequal distribution of power and opportunities (Hartmann and Boyce, 1983; Hewitt, 2007). In cases of extreme deprivation, homeless people, migrants, travelling communities, squatters, gender minorities, lower untouchable castes and other social groups whose citizenship and identity are legally or socially illegitimate (or both) or not properly acknowledged through state identification systems cannot even claim such access. This constitutes a further obstacle to protection and driver of vulnerability. Therefore, the vulnerability–marginality nexus here goes much beyond poverty and includes issues of age, physical ability, gender, sexuality, race, ethnicity, caste and religion (Wisner et al., 2012).

People's vulnerability to natural hazards varies in time and space and mirrors the nature, strength and diversity of livelihoods. The intimate relationship between livelihood and vulnerability explains why many people frequently have no other choice but to face natural hazards to sustain their daily needs (Maskrey, 1989; Wisner, 1993). In many instances, threats to everyday needs, especially to food security, are almost always more pressing than threats from rare or seasonal natural hazards. The difficulty of meeting daily needs can lead to environmental degradation, which often manifests itself in increasing natural hazards (Lewis, 1999; Wisner et al., 2004).

For the proponents of the vulnerability paradigm, people's vulnerability to natural hazards therefore results from their limited ability to control their daily lives. In that context, disasters highlight or amplify daily hardship and everyday emergencies associated with food insecurity, illnesses, fragile shelter and poverty at large. Therefore, disasters cannot be considered accidents beyond the usual functioning of the society (Hewitt, 1983). Instead, disasters generally reflect development failures where the root causes of vulnerability have origins in other, usually contextual, development-related crises. The root causes of disasters are grounded in an array of structural socio-cultural heritages and political economy processes working at both local and international scales. These are further explained by deep-seated historical processes, such as colonial and neocolonial legacies (Watts and Bohle, 1993). Ultimately, the root causes of disasters most often lie beyond the hands of those who are vulnerable and suffer.

Understanding people's vulnerability requires fine-grained study which relies on qualitative and often ethnographic research methods through undertaking what Chambers (2007a) has called anthropological particularism. If the outcomes of the growing number of studies conducted under the vulnerability paradigm over the past three decades are not directly comparable, some powerful frameworks have been designed to confront them. The most influential of these is the Pressure and Release (PAR) framework devised by Davis (1978, 1984) but popularized by Blaikie et al. (1994). The PAR tracks down the factors of vulnerability from the unsafe everyday lives of those at risk to dynamic pressures and, ultimately, structural root causes.

The foregoing tenets of the vulnerability paradigm constitute a radical departure from previous interpretations of disasters. It has been criticised for putting too much emphasis on the social to the detriment of the natural component of disasters. In addition, many scholars working in wealthy Western countries have found it more applicable to the less affluent regions of the world where this approach emerged in the first place. In its early years, the vulnerability paradigm was also challenged for overly stressing people's weaknesses to the detriment of their intrinsic capacities to face natural hazards and disasters.

From vulnerability and suffering to capacities and resilience

Labelling some people 'vulnerable' often leads to their being stigmatised as 'helpless victims' of an unjust society and more powerful individuals. There has been increasing recognition that this is not a true depiction of the response of people to hazards and disasters. Throughout the 1970s and 1980s, practitioners of disaster risk reduction at the local level, especially in Latin America, South Asia and Southeast Asia, observed that people are often proactive and creative in facing hazards and disasters (e.g. Cuny, 1983; Delica, 1993; Maskrey, 1989; Wisner et al., 1977). As Anderson and Woodrow (1991, p. 47) put it, 'no matter how poor nor how much they have lost through a disaster, people still have some material capacities'. People always display some form of ability to face natural hazards and disasters, which is gathered under the concepts of both capacities and resilience.

Both capacities and resilience build on evidence that people are the first responders in dealing with hazards and disasters (Delica-Willison and Gaillard, 2012; Maskrey, 1984, 1989). People rely on kinship ties and solidarity networks, remittances, traditional medicines, vernacular architectures, experiences and knowledge of past hazardous events, traditional water management, hazard-resistant crops, the knowledge of biodiversity resources, the ability to hunt and fish, local leadership, and flexible decision-making processes in dealing with hazardous phenomena in their environment. In times of disaster, outside assistance arrives, at best, hours or at least days after the event, even though it is well known that the initial few hours are crucial to save lives and livelihoods. Evidence collected by sociologists actually suggests that a very large majority of post-disaster survivors are rescued by their friends, kin or neighbours who are on the spot at the time of an event and who rely upon local knowledge and available resources and skills (Quarantelli and Dynes, 1972).

Although the increasing emphasis given to people's capacities and resilience emerged as a spin-off to the vulnerability approach and has not led to the emergence of a new paradigm per se, it still marks a significant evolution in the study and understanding of disasters. The concept of capacities has been widely used through the practice of Vulnerability and Capacity Analysis (VCA), a core dimension of many disaster risk reduction projects derived from Anderson and Woodrow's (1989) framework for assessing vulnerability and capacities. Meanwhile, the concept of resilience has been broadly popularised by scholars throughout the 1980s and 1990s and has since percolated into policy and practice to become one of the most widely used concepts in the field. Because they are often embedded in people's everyday lives and cultures, capacities are difficult to apprehend for outside researchers and practitioners. For this reason, the analysis of capacities has largely relied upon participatory approaches led by those at risk along the line of what Chambers (2007a) calls participatory pluralism, which includes tools, methods, attitudes and behaviours geared towards fostering the contribution of a large array of diverse and frequently excluded local actors in analysing and finding solutions to the problems that affect them. On the other hand, the assessment of resilience follows multiple, sometimes antithetical approaches, which we explore in chapter 4.

The recognition of people's capacities in dealing with disasters contributed to the broader consideration of local people's participation in development. The emergence of the concepts of capacities in both disaster studies and policies and also in actions to reduce disaster risk was strongly influenced by the growing and concomitant momentum gained by the idea that people, including the poor and those who face natural hazards, should be at the forefront of development because they are knowledgeable and resourceful (Chambers, 1983; Freire, 1968; Hall, 1978). Recognising that people have capacities underpins the assertion that they should also participate in disaster risk reduction or, as put in many Red Cross and Red Crescent Societies manuals in the early 2000s, that they should 'have more control over shaping their own futures' (e.g. Vietnam Red Cross Society, 2000, p. 6). This suggests a shift in power relations to the detriment of outside institutions and organisations who, at the time, were considered the dominant, if not exclusive, stakeholders of disaster risk reduction. Capacities, in fact, emerged alongside the concept of empowerment (Chapter 7).

The political agenda that emerged with the concept of capacities, however, has largely vanished with the concurrent rise of the concept of resilience and its 'mainstreaming' into a buzzword; one that is associated with the rise of neoliberalism and its project to reaffirm the ideals of the Enlightenment and the individual's right to freedom and free thought (Bankoff, 2019; Rigg and Oven, 2015). Resilience emphasises the ability of individuals to take the lead in reducing risk, thus providing an opportunity for government agencies to withdraw both investment in disaster risk reduction and social protection. As Harvey (2005, p. 23) writes, with neoliberalism, 'all forms of social solidarity [i.e. associated with the State] were to be dissolved in favour of individualism, private property, personal responsibility, and family values'. The progressive and continuing withdrawal of the state in disaster risk reduction, associated with both neoliberalism and the rise of resilience, has concurrently fuelled the growth of the civil society organisations, especially non-governmental organisations (NGOs), whose role has been to replace that of a vanishing state. Donor organisations, in all their diversity, have promptly jumped onto such an opportunity to promote a humanitarian agenda that shifts the responsibility away from the state (Prashad, 1993). As a result, the concept of resilience has inadvertently led to a broader discourse antithetical to its original ethos. In fact, its very diverse interpretations have come in support of all paradigms previously discussed (Chapter 4).

Dialectical tradition and the Western legacy

Concepts and paradigms in disaster studies have been the building blocks for broader discourses on disasters. These have penetrated the broader space of popular culture and disaster risk reduction policies and actions, including the *dispositif* that supports them. They have trickled through multiple channels, including movies, literature, traditional and social media, international agreements and national policies, and through diverse institutions such as international organisations, government agencies, NGOs and schools. Although the hazard paradigm is still prominent in popular culture and the media (both traditional and social), attention has largely shifted to vulnerability and then further to capacities/resilience among international organisations, NGOs and many governments throughout the world. We explore these discourses further in Chapters 5, 6 and 7 of this book.

At this point, therefore, one may perceive an intriguing co-existence of different discourses that seem incoherent at first sight. However, it is our contention that this seeming incoherence is artificial if one considers the academic debate between the proponents of the hazard and vulnerability paradigms within the realm of the broader modern *épistémè* of the West that envelops all these approaches. Indeed, all of the concepts upon which the different paradigms build are of Western origins or reflect a Western interpretation of an Arabic term in the case of hazard. As such, they mirror a single academic heritage, that of the West and the Enlightenment.

The Western academic heritage is obviously not monolithic. The different paradigms available in disaster studies reflect different epistemologies that often are conflicting. The positivist, constructivist and pluralist approaches to understanding disasters respectively promoted by the hazard and vulnerability paradigms, as well as the capacities/resilience spin-off approach of the latter, are neither homogenous. They reflect different perspectives inspired by diverse academic traditions and methodologies that we discuss in Chapter 4. However, these paradigms and their very dialogue over the past century can be firmly grounded and reduced to the dialectical nature of the nature–culture binary.

In fact, one may look at the successive emergence of the hazard and vulnerability paradigms in disaster studies as a form of Hegelian dialectic, a historical journey towards progress and reason. The ambition of the hazard paradigm was to provide the ultimate truth of the time. As such, it was to support a standardised approach to understanding people's behaviour in dealing with disaster and hence to bringing progress to society by significantly reducing the risk of disaster through the transfer of experience and technology. The subsequent emergence of the vulnerability paradigm in the 1970s was a direct and explicit response to the then-dominant hazard paradigm. Many of the early seminal texts from the proponents of the vulnerability paradigm were, in fact, critical reviews of the shortcoming of the hazard paradigm (Hewitt, 1980; Torry, 1979b; Waddell, 1977).

The organically dialectical nature of disaster studies to date may be broken down a level further within each main paradigm. The human ecology/

behaviourist dimension of the hazard paradigm is a response to the previous view that the hazard was the only dimension of disasters worthy of interest in the perspective of reducing disaster risk. The human ecology/behaviourist approach introduced a human dimension to the study of disaster yet continued to place hazards, in their rare and extreme occurrence, as the main driver of people's behaviour. Similarly, the emergence of the concepts of capacities/resilience in disaster studies pushed back against the view that people are solely vulnerable in dealing with hazards and disasters while still aligning with many of the tenets of the vulnerability paradigm when it comes to the importance of unpacking power relations to understand disasters.

Scholars have recently suggested that a third perspective is beginning to bridge the gap between the hazard and vulnerability paradigms (e.g. Hilhorst, 2004; McEntire, 2001), thus suggesting the onset of the third step in Hegel's (1807) classic pattern of dialectic. This possible third paradigm, called holistic by McEntire (2001) and mutuality or complexity paradigm by Hilhorst (2004), the contours and ethos of which are still to be refined if not defined in the first place, aims at understanding the interactions between nature and society through complexity theory. Others (e.g. Turner II et al., 2003) are taking hazard-centred and apolitical approaches to vulnerability. These consider the demographic profile of people such as their gender, age and ethnicity as factors of vulnerability. This approach overlooks power relations and often takes a taxonomic perspective in direct relation to hazards through the bridging concept of exposure. We provide a critique of these emerging approaches to the concept of vulnerability in the next chapter.

The intrinsically dialectical nature of disaster studies mirrors a continuing quest for reason and progress, firmly grounded in the modern *épistémè* of the West and the legacy of the Enlightenment. It has been possible because of the dialectical nature of the ontological assumption that disasters sit at the interface between nature/hazard and culture/vulnerability; a divide we have seen that is at the core of the modern *épistémè* of the West and the project of modernity inherited from the Enlightenment (Horkheimer and Adorno, 1947; Smith, 2008). The progressive shift of attention from hazard/nature to vulnerability/culture may seem like a radical turn in our understanding of disasters if taken from within a Western perspective. However, if one steps back further outside of the underpinning divide between nature and culture, it looks like a mere shift of the centre from one side to the other within the same binary. A shift which is neither challenging the underlying ontological assumption inherited from the Enlightenment nor the very conjecture that there is such a thing as a single truth with regard to the nature of what we call disasters as well as to their causes.

Conclusion

Such ontological and epistemological landscapes for the emergence, development and strengthening of disaster studies have two major implications for the argument put forward in this book, which is that the hegemony of

Western understandings of disaster makes it possible for standardised and normative policies and actions for reducing disaster risk that overlook the very diverse realities and world views/senses of those who are affected by what we call disaster.

First, it sets Western ontologies and epistemologies as the one and only truth when it comes to both defining and explaining disasters. The primacy of reason is the direct and hegemonic legacy of the Enlightenment and alleged prominence of Western science, one that Husserl (1954, p. 326) called *theoria* in comparing Western (i.e. inherited from the Greeks) and non-Western philosophies:

> Aber nur bei den Griechen haben wir ein universales ("kosmologis-ches") Lebensinteresse in der wesentlich neuartigen Gestalt einer rein "theoretischen" Einstellung, und als Gemeinschaftsform, in der es sich aus inneren Gründen auswirkt, die entsprechende wesentlich neuartige der Philosophen, der Wissenschaftler (der Mathematiker, der Astronomen usw.). Es sind die Männer, die nicht vereinzelt son-dern miteinander und füreinander, also in interpersonal verbundener Gemeinschaftsarbeit, Theoria und nichts als Theoria erstreben und erwirken, deren Wachstum und stetige Vervollkommnung mit der Verbreitung des Kreises der Mitarbeitenden und der Abfolge der Forschergenerationen schließlich in den Willen aufgenommen wird mit dem Sinn einer unendlichen und allgemeinsamen Aufgabe. Die theoretische Einstellung hat bei den Griechen ihren historischen Ursprung.[6]

Second, the prominence of Western ontologies and epistemologies provides ground for a binary understanding of the world, which sets a safe West apart from a dangerous rest of the world (Bankoff, 2001). This imaginary representation of the world through a Western gaze mirrors Said's (Said, 1978, pp. 45–46) analysis of the *Orientalisation* of the Middle and Far East throughout the past three centuries:

> When one uses categories like Oriental and Western as both the starting and the end points of analysis, research, public policy (as the categories were used by Balfour and Cromer), the result is usually to polarize the distinction-the Oriental becomes more Oriental, the Westerner more Western-and limit the human encounter between different cultures, traditions, and societies. In short, from its earliest modern history to the present, Orientalism as a form of thought for dealing with the foreign has typically shown the altogether regrettable tendency of any knowledge based on such hard-and-fast distinctions as "East" and "West": to channel thought into a West or an East compartment. Because this tendency is right at the center of Orientalist theory, practice, and values found in the West, the sense of Western power over the Orient is taken for granted as having the status of scientific truth.

This deep belief that Western understandings of disaster constitute the one and only truth and the concomitant assumption that the West knows best how to deal with disasters have justified and supported the imperialist ideology that characterises the disaster risk reduction policies and actions, in all their diversities, deployed around the world since the 1970s. Althusser's (2018) nexus between knowledge, ideology and hegemony is here complete. Progress towards modernity driven by reason, as fostered and experienced by the West since the late eighteenth century, hence needs to be brought to the dangerous rest of the world so that their suffering is ended. However, today, 'it is the "rationalism" of these ideologies of progress that increasingly comes to be eroded in the encounter with the contingency of cultural difference' (Bhabha, 1994, p. 280).

It is then the very possibility for multiple *épistémès* to co-exist that is at stake. If Foucault (1966) refuted this possibility in the first place, it was from the perspective of Western philosophy. Postcolonial studies, in the path of Said, Spivak and Bhabha, have clearly opened up further opportunities for thinking beyond this narrow and monolithic perspective. In disaster studies, these opportunities require us to (re-)examine the concepts we use, our ways of knowing, and the ways in which we design our policies and actions towards reducing disaster risk, to challenge both the universal dimension of Western ontologies and epistemologies and the existence of other world views/senses and ways of knowing.

Notes

1 '(...) *admit, for example, that nature did not construct twenty thousand houses of six to seven stories there, and that if the inhabitants of this great city had been more equally spread out and more lightly lodged, the damage would have been much less, and perhaps of no account. All of them would have fled at the first disturbance, and the next day they would have been seen twenty leagues from there, as gay as if nothing had happened; but it is necessary to remain, to be obstinate about some hovels, to expose oneself to new quakes, because what is left behind is worth more than what can be brought along. How many unfortunate people have perished in this disaster because of one wanting to take his clothes, another his papers, another his money!'* From the English edition translated by Judith Rush, Roger Masters, Christopher Kelly and Terence Marshall and published by the University Press of New England in 1992.

2 '*The associations of men are in great part the work of accidents of nature; particular floods, overflowing seas, volcanic eruptions, great earthquakes, fires kindled by lightning and which destroyed forests, everything that must have frightened and dispersed the savage inhabitants of a land must thereafter bring them together to repair in common their common losses. The traditions of the earthly calamities so current in ancient times show what instruments providence used to force human beings to come together. Ever since societies have been established these great accidents have ceased and become more rare; it seems that this too must be so; the same calamities that brought together scattered men would disperse those who are united*'. From the English edition translated by John Scott and published by the University Press of New England in 1998.

3 '*The doubling of nature into appearance and essence, effect and force, made possible by myth no less than by science, springs from human fear, the expression of which becomes its explanation*'. From the English edition translated by Edmund Jephcott and published by the Stanford University Press in 2002.

4 '*Absolute purpose of life*'. From the English edition translated by Edmund Jephcott and published by the Stanford University Press in 2002.

5 '*Events, to indicate that they occur without a necessary or expected cause*'. Our translation.

6 '*But only in the Greeks do we have a universal ("cosmological") life-interest in the essentially new form of a purely "theoretical" attitude, and this as a communal form in which this interest works itself out for internal reasons, being the corresponding, essentially new [community] of philosophers, of scientists (mathematicians, astronomers, etc.). These are men who, not in isolation but with one another and for one another, i.e., in interpersonally bound communal work, strive for and bring about theoria and nothing but theoria, whose growth and constant perfection, with the broadening of the circle of coworkers and the succession of the generations of inquirers, is finally taken up into the will with the sense of an infinite and common task*'. From the English edition translated by David Carr and published by the Northwestern University Press in 1970.

3 Unfulfilled promise of a paradigm shift

In this chapter, we focus on how the attention of scholars of disasters has moved from nature/hazard to culture/vulnerability over the past 50 years. This paradigm shift, which we briefly discussed in the previous chapter, has been seen as a major step forward. This step has allowed a better understanding of disasters and hence a positive refinement of policies and actions to reduce disaster risk (see Chapter 5). We have (almost) all come to rebut the proposition that disasters are natural, so that anyone daring to use the misnomer 'natural disaster' now sparks widespread criticism. We have all come to recognise that disasters result from the unequal distribution of power and resources between those who are more vulnerable and those who are less so. Yet, in saying this, we have also recognised that even those most vulnerable are not helpless victims when dealing with disasters. They all possess knowledge, skills and resources that gather as capacities.

In pushing for this paradigm shift, we have all claimed to be innovative and critical, so that the vulnerability paradigm is also known as the radical paradigm and therefore is an apologia for critical scholarship. As such, disaster studies has been instrumental in the emergence of broader and highly influential fields of scholarship, such as political ecology and environmental justice, among others. But, as Jim Blaut (1993) once realised in the context of Western imperialism, we may be facing a fascinating anomaly. Have we indeed grasped the full implications of our critique of the hazard paradigm, or are we perpetuating some of its core (and most problematic) tenets? Have we actually taken on the challenge set up for us 50 years ago by the pioneers of the vulnerability paradigm?

This chapter argues that we may have only partially done so. In fact, in many aspects, disaster studies still mirrors a Western hegemony that we were meant to challenge in the first place. In the previous chapter, we discussed that, in the end, this alleged radical paradigm shift has been, in practice, a mere dialectical movement across the hazard–vulnerability binary and that, as a result, disaster scholarship has remained firmly grounded in the assumption of the West, inherited from the Enlightenment, that disasters sit at the interface between Nature and Culture. We contend here that our inability to fulfil the promise of a truly radical agenda, at its inception, to step outside of Eurocentric/Western ontologies and epistemologies

DOI: 10.4324/9781315752167-3

in understanding disasters, reflects unequal power relations between researchers and research institutions located in the West and those, in all their diversity, who live in other regions of the world.

A brief epistemology of a paradigm shift

Let us recall that the vulnerability paradigm emerged in the 1970s in reaction to the then-dominant hazard paradigm. The vulnerability paradigm was then designed to push back against the idea that disasters are the consequence of *extra*-ordinary hazards that overwhelm people and societies (Hewitt, 1983). Rather, it encouraged us to consider disasters within the context of everyday life and the way power and resources are shared within society – that is, to appraise vulnerability to disaster as a cultural, social, political and economic construct. This was seen as a critical departure from the then-common under-standing of disasters. We had to 'radically rethink the causal relationships involving people and nature' (Wisner et al., 1976, p. 548).

The vulnerability paradigm was the bearer of a strong political agenda amid emerging and broader postcolonial thought in the paths of Aimé Césaire (1950, 1956), Frantz Fanon (1952, 1961), Paulo Freire (1968), Fabien Eboussi-Boulaga (1977) and Edward Said (1978). The unequal power relations between, on the one hand, those who had long defined international scholarship in Europe and other Western countries and, on the other hand, those who had studied in Asia, Africa and Latin America, was at the core of this agenda. Western technocratic views of disasters were considered skewed and inappropriate inside their homelands and more significantly outside. They were seen as the mere justification for imposing neocolonial policies and actions to reduce the risk of disaster in the rest of the world (Comité d'Information Sahel, 1975; O'Keefe et al., 1976; Wisner et al., 1976).

'A change in the whole approach to disaster' was needed (Lewis, 1976a, p. 8). We were challenged not only to amend the way we understood disas-ters, moving from nature to society, but also to reconsider the way we study them and the way we come to think about them (Ball, 1975; Copans, 1975; O'Keefe et al., 1976; Wisner et al., 1977). As Waddell (1977, pp. 75–76) then suggested, our interpretation of disasters was 'dictated by the con-straints of the methodology' that was not 'necessarily dictated by reality, but rather by a social scientific tradition in the West which fragments reality and which promotes a type of functional analysis that is profoundly ahis-torical'. We therefore had to move away from rigid research methods, such as relying on standardised questionnaires that were designed by outsiders who are Westerners. Chambers (1981) called these approaches 'quick-and-dirty' and criticised them for being skewed by the so-called tarmac and dry-season biases where outside researchers focus primarily on easily acces-sible places at convenient times of the year (see Chapter 4).

Therefore, we were encouraged to embark on an epistemological journey that was meant to take us away from the certainties of Western scholarship.

We were meant to challenge the hegemonic rules and values of Western science that were underpinning the whole transfer of knowledge and technology associated with the then-dominant strategies to reduce the risk of disaster. These strategies were embedded within the broader imperialist and neocolonial relationships imposed by Western governments upon the rest of the world (Comité d'Information Sahel, 1975; Copans, 1975).

The concept of vulnerability was the springboard for this journey but was in no way a silver bullet. Rather, it was used as a prompt to uncover issues and processes that lead people to be adversely affected in the event of hazardous phenomena. Lewis (1979, p. 116) was even questioning whether the very concept of disaster could be 'a wholly Western concept, introduced by alien administrations from alien sources and adopted for practical and pragmatic advantages?' As Young (2004, p. 46) further argued,

> (…) by definition the concept 'cannot capture the absolutely-other'; and, to the extent that it must invoke a form of generality, of language itself. Any conventional form of understanding must appropriate the other, in an act of violence and reduction.

In no way, therefore, were Western concepts meant to be rolled out in all sorts of settings and locations as the panacea to understand and address the root causes of people's hardship (Richards, 1975). This would be in contradiction to the essence of the paradigm shift and to the very idea that disasters are social constructs.

Research, rather, was meant to be driven by local scholars within their own countries (Lewis, 1979) or by local people themselves through genuine participatory research outside of the academic environment (Wisner et al., 1977). Local researchers were meant to study disasters on their own terms through local and indigenous perspectives and concepts. Research thus was to be moved away from the silo of Western science and academic institutions, whose role, beyond their surrounding localities, was supposed to shift from drivers to supporters. We were all to acknowledge that local researchers and people affected by disasters are as good and capable as Western scholars and their views could underpin indigenous and context-specific initiatives to reduce the risk of disaster as well as support their demand for action from the state.

Have we risen to the challenge?

Almost 50 years on, have we really completed this 'revolution in thinking about disasters' (Wisner et al., 1976, p. 548) initiated in the 1970s? For sure, as discussed in the introduction of this essay, our general understanding has changed to better capture the social dimensions of what we call disasters.

The concept of vulnerability has become a mainstay of disaster research, and virtually all researchers interested in studying disasters, whatever their background, are now handling the concept in one way or another (Hewitt,

1995b; Wisner, 2016). Many have made it their own, sometimes with very meaningful perspectives, sometimes within a taxonomic approach and, in other instances, in direct link with natural hazards, through cognate concepts such as exposure (to natural hazards) and indicators like demographic data. In many instances, therefore, vulnerability has been emptied of its political and social essence. The political agenda has often vanished.

Both concepts of vulnerability and disaster have also been rolled out across continents, including in places where they cannot be translated into local languages. The concept of vulnerability, in particular, has been imposed upon people who have been struggling to make sense of its scope, as if adopting the language of the West was a symbol of elevated status and more rigorous values (Fanon, 1952, 1956; Eboussi-Boulaga, 1977; Salazar, 1991). A couple of decades ago, in a seminal, unconventional article, Mihir Bhatt (1998) actually asked whether vulnerability could – and should – be understood beyond its Western academic acceptance. Nonetheless, many studies are still framed through the lens of Western ontologies and epistemologies, perpetuating a hegemony that was meant to be challenged. As Mihir Bhatt (1998, p. 71) suggests, studies driven by an outside researcher are likely to be 'filtering what she or he reads through the conceptual framework, assumptions, and values or her or his culture and, as a result, is creating false 'stories' that fit her or his expectations'.

Ultimately, disaster studies thus continues to be dominated by Western scholars, whatever the location of the disasters or study areas. A review of articles published in the journal *Disasters* since its inception in 1977 shows that 84% of authors are affiliated with institutions based in countries that are members of the Organisation for Economic Co-operation and Development (OECD) (as an imperfect but probably closest available proxy for the West). In the meantime, 93% of those who died in large disasters were living in non-OECD countries, according to the Emergency Events Database (EM-DAT)[1] of the Centre for Research on the Epidemiology of Disasters (2018).

A closer examination of the broader Anglophone literature (yet another Western bias) for those that have stirred the largest number of publications further mirrors the Western hegemony over disaster studies (Figure 3.1). American, New Zealand, Japanese and Chinese (for the sake of the exception amidst OECD member countries) researchers have largely controlled research carried out in their own countries following recent major disasters. On the other hand, most of the research initiatives conducted in the aftermath of disasters that occurred in the Philippines, Nepal and Haiti have been led by scholars based in OECD countries. In such a context, how can local knowledge be considered paramount and local stakeholders be the leaders of disaster research? Isn't there enough capacity elsewhere than in the West (and China) to conduct research? Aren't we reproducing a pattern that Hewitt (1983, p. 14) flagged as characteristic of the hazard paradigm that we were meant to challenge:

It (i.e. hazards research) does not reflect upon the flaws in itself, expect in relation to what is deemed sophisticated in the current fashions of the scientific community. It gathers data about people at risk, but may not engage in a dialogue with them. Most disaster reports in the so-called Third World are by persons who cannot speak the language of the area affected, or have no background in its sociocultural composition.

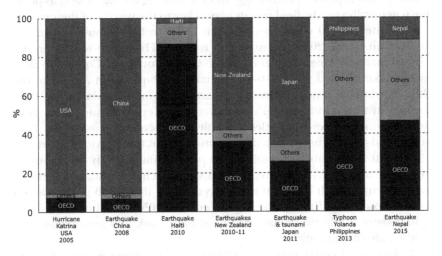

Figure 3.1 Distribution of authorship (based on affiliation of lead authors) for the seven disasters that stirred the greatest interest between 2005 and 2015 (data from Scopus)

Furthermore, researchers continue to prioritise large disasters to the detriment of smaller events despite the mounting evidence that the latter are the priority of the most people, mainly because small events have a larger cumulated impact and contribute to a ratchet process of deprivation among those affected (Lavell and Maskrey, 2014). The continuing interest of Western scholars in large disasters reflects a 'gold rush' approach to research.

On the disaster 'gold rush'

Dynes et al. (1967, p. 46) once set the agenda for disaster research as follows:

High priority is given to those disasters which are quick and unexpected, which affect more than one industrial community, where there is heavy property damage, where the number of casualties exceeds 100 and which elicits the participation of national organizations during the emergency period.

This was no new priority. The 1631 eruption of Mt Vesuvius is believed to have triggered an 'epidemic' of research (Scacchi, 1882, p. 1): *'l'incendio del*

1631 produsse, mi si permetta il dirlo, una epidemia tipografica, per cui nomini culti ed ignoranti si affrettarono più che a descrivere a commemorare la sterminatrice con-flagrazione[2]. Almost 400 years on, major disasters continue to stir the prime interest of researchers who frequently rush immediately to the affected areas to conduct studies of various kinds, from hazard observations to social surveys on the impact of the events and to post-traumatic stress disorder research. Stallings (2007, p. 56) actually suggests that 'arriving on site as soon as possible is generally seen by field researchers as key to the success of their work'.

Over the past 20 years, this research 'gold rush' was observed in the regions hit by the 2004 Indian Ocean tsunami, Hurricane Katrina in the United States of America in 2005, the 2008 earthquake in China, the 2010 earthquake in Haiti, the 2010–11 Ōtautahi earthquakes in Aotearoa, the 2011 earthquake and tsunami in Japan, Typhoon Yolanda in the Philippines in 2013, and the 2015 earthquake in Nepal. A quick analysis of academic peer-reviewed articles available from Scopus (those which have stimulated the highest academic attention over the previous 20 years) shows that the number of publications peaked immediately or, at most, a year after these events (Figure 3.2). This is particularly evident for Hurricane Katrina, which has been the focus of more than 5500 peer-reviewed publications, including 424 before the end of 2005. Not all of these quick post-disaster publications have required field work and immediate field studies, although many have.

Rushing to affected areas immediately after the event is very tempting for researchers interested in disasters. What White and Haas (1975) called '*postaudits*' have long been deemed essential for better understanding the impact of natural hazards as well as the response of people to the events

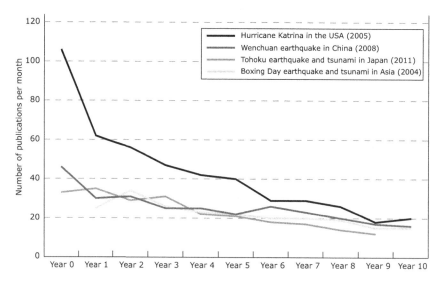

Figure 3.2 Number of publications per month after the four disasters that stirred the highest interest between 2004 and 2020 (data from Scopus)

and, as a result, for enhancing policies for disaster risk reduction (Killian, 1956; Mileti, 1987; Stallings, 2002, 2007). Quarantelli (1997, p. 57) provides two basic reasons why it is allegedly so important to get to the scene as soon as possible after the event:

> First, observations can be made and documents can be collected that cannot be obtained through later interviewing. The social barriers that normally exist to restrict access to high level officials and key organizations, simply do not exist. Second, being on the scene early insures a high degree of access and cooperation. Victims are typically candid, cooperative and willing to talk in ways far more difficult to get later.

Stallings (2007, p. 61) adds that eventually 'respondents' personal recall that may be skewed by repeated retelling of their stories to a succession of interviewers'. Researchers who rush to disaster-affected areas thus justify their approach by the perishable nature of the data they need to collect (Bourque et al., 1997).

Although the collection of perishable data may be considered important (both for the sake of the region affected and for advancing science), the multiplication of initiatives from different countries and research groups leads a very large number of individuals to converge on places affected by what we call disasters. For example, in the immediate aftermath of the 2004 tsunami, teams of physical and social scientists from France, Japan, Russia and the United States of America, among others, collected data in Indonesia with little or no coordination at first (e.g. Borrero, 2005; Kawata et al., 2005; Iemura et al., 2006; Moore et al., 2006; Lavigne and Paris, 2011), so that some suggested that Aceh was actually 'ransacked' by researchers (Missbach, 2011). Moreover, the disaster 'gold rush' often (though not always) sees the convergence of researchers with limited knowledge of the area and insufficient time to conduct a prior appropriate literature review, an issue that was underlined by Killian (1956) 60 years ago and that can lead to misconceptions (Gomez and Hart, 2013).

We recognise that post-disaster research has emerged from a real desire to 'do good'. The 'gold rush' is usually motivated by a genuine will to contribute to the recovery and future disaster risk reduction effort by enhancing our understanding of the event or process. Again, this is a fair and commendable objective for researchers familiar with both the place and the local culture because their pre-existing networks of stakeholders and friends, including those directly affected, may allow them to collaborate in post-disaster recovery or disaster risk reduction in the continuity of their previous engagements.

Nonetheless, one may wonder whether it is appropriate for outsiders less familiar with the affected places, who may lack prior cultural and language skills, to converge on places where people struggle to rebuild their lives and livelihoods and where they might have priorities other than answering questions about their recent experiences (Schenk, 2013). Ethically, these outside researchers may sit on the fence, as suggested in the larger field of development studies (e.g. Sidaway, 1992; Cooke, 2004). If we accept behaviours that otherwise would not be allowed following disasters, the controversial actions are all answering an imperative of immediate necessity and immediate 'greater good' for affected places and local people.

However, outside researchers do not always contribute to the greater good. If one considers that the ultimate goal is the rebuilding of the place and of lives and livelihoods (i.e. to the greater good or at least the good for the greater number), then the presence and actions of outside researchers can be assessed in term of impacts to the local people, regardless of the researchers' own objectives and motivations. One may then ask how researchers who need a translator, who extract data from those affected by disasters, who possibly behave inadequately because of a lack of cultural understanding, and who also consume local resources to sustain their own needs (which frequently differs from that of the people affected by the event) have a positive effect on the local rebuilding[3].

These questions have long been asked in the broader literature on the ethics of academic fieldwork, especially in the context of researchers from wealthy countries visiting less affluent countries (e.g. Desai and Potter, 2006; Scheyvens and Storey, 2014). However, in a post-disaster context, all of these issues are exacerbated. In such a case, the presence of outside researchers shortly after a disaster might be inappropriate, as they may be detrimental to the good of the greater number.

Perpetuating the Western hegemony in disaster studies

In the end, it turns out that the way disaster studies are conducted has not really changed over the past 40 years, certainly not as much as the pioneers of the vulnerability paradigm envisioned back in the 1970s. Disaster studies, therefore, very much mirrors Oyěwùmí's (1997, p. 18) bleak assessment of contemporary research on gender, particularly in the African context:

> The questions that inform research are developed in the West, and the operative theories and concepts are derived from Western experiences. African experiences rarely inform theory in any field of study; at best such experiences are exceptionalized. Consequently, African studies continues to be "Westocentric", a term that reaches beyond "Eurocentric" to include North America.

In fact, there seems to be more rhetoric than actual commitment to change and one may agree with Butler (2016, p. 22) that 'one political problem that emerges from any such discussion is whether the discourse on vulnerability shores up paternalistic power, relegating the condition of vulnerability to those who suffer discrimination, exploitation, or violence'. In other terms, studies of disasters may be an instrument of control, one that supports the hegemony of Western concepts and discourses in disaster studies and disaster risk reduction.

Hegemony is a soft and subtle approach to exerting power based on consent rather than coercion (Gramsci, 1971). It mirrors 'political leadership based on the consent of the led, a consent which is secured by the diffusion and popularisation of the world view of the ruling class' (Bates, 1975, p. 352). In Gramsci's perspective, intellectuals, including researchers of all types as well as journalists who spread their word across society, are key to generate and convey the knowledge that underpins the world view of those in power (Gramsci, 1930; Femia, 1981). In fact, concepts and discourses developed by Western scholars are central to understanding the perpetuation of the hegemony of the West in disaster studies.

The constant reiteration of these concepts through multiple channels (science, policy, media, infrastructure, etc.) is central to the construction of broader discourses (Butler, 1993) and leads them to be taken for granted. As such, they become the one and only truth when it comes to studying disasters. Gramsci (1971) and Fanon (1952) refer to this acceptance as common sense so that those who are dominated – the subalterns – do not think of challenging the ideas (and the associated ideology) of those in power. Said adds (1994, p. 15) that 'the power to narrate, or to block other narratives from forming and emerging, is very important to culture and imperialism, and constitutes one of the main connections between them'.

This negative conditioning of those upon whom such concepts and discourses are imposed contributes to the 'mystification', to use a word introduced by both Fanon (1956) and Memmi (1957), of non-Western ontologies and epistemologies but also of non-Western scholars. As Fanon (1952) and Freire (1968) would argue, scholars from the non-Western world actually embrace such concepts and discourses in the hope of stepping up to the level of scientific and political prominence of those who generate and spread them, the Western researchers. As a result, they consent to the exclusion of other forms of ontologies and epistemologies to understanding disasters. This consent ultimately ensures the strength and durability of Western discourses and their hegemony in disaster studies as in other fields (Said, 1978). As Oyěwùmí (1997, p. 16) once stated,

> Academics have become one of the most effective international hegemonizing forces, producing not homogenous social experiences but a homogeny of hegemonic forces. Western theories become tools of hegemony as they are applied universally, on the assumption that Western experiences define the human.

Again, our intention is not to challenge the relevance of these concepts and discourses within the West. Our concern is their rolling-out around the world in places where they do not translate or make sense (Chmutina et al., 2020). When applied to the non-Western world, concepts such as disaster and vulnerability contribute to a skewed and homogenous representation of very diverse local realities. Ironically, Lewis stated, back in 1979 (pp. 113–114), that

preoccupation with Western concepts and Western disasters and Western outsider response to overseas disaster has hindered any study and
analysis of the perception of and response to hazard in third-world
countries, and in societies and cultures different from our own.

The hegemonic use of these concepts and surrounding discourses contributes
to the fabrication of an artificial divide between a safe West, that suffers least
from disasters, and a dangerous rest of the world, that suffers most (Bankoff,
2001). The rest of the world includes Said's (1978, pp. 21–22) Orient:

> Another reason for insisting upon exteriority is that I believe it needs to
> be made clear about cultural discourse and exchange within a culture
> that what is commonly circulated by it is not "truth" but representa-
> tions. It hardly needs to be demonstrated again that language itself is a
> highly organized and encoded system, which employs many devices to
> express, indicate, exchange messages and information, represent, and
> so forth. In any instance of at least written language, there is no such
> thing as a delivered presence, but a re-presence, or a representation.
> The value, efficacy, strength, apparent veracity of a written statement
> about the Orient therefore relies very little, and cannot instrumentally
> depend, on the Orient as such. On the contrary, the written statement
> is a presence to the reader by virtue of its having excluded, displaced,
> made supererogatory any such rear thing as "the Orient". Thus all of
> Orientalism stands forth and away from the Orient: that Orientalism
> makes sense at all depends more on the West than on the Orient, and
> this sense is directly indebted to various Western techniques of rep-
> resentation that make the Orient visible, clear, "there" in discourse
> about it. And these representations rely upon institutions, traditions,
> conventions, agreed-upon codes of understanding for their effects, not
> upon a distant and amorphous Orient.

Alleged differences between the West and the rest of the World were at the
centre of the colonial agenda and these continue to support neocolonial
and imperialist ideologies (Horkheimer and Adorno, 1947; Fanon, 1956;
Memmi, 1957). Asad (1973, pp. 18–19) once suggested that

> it is essential to turn to the historical power relationship between the
> West and the Third World and to examine the ways in which it has been
> dialectically linked to the practical conditions, the working assump-
> tions and the intellectual product of all disciplines representing the
> European understanding of non-European humanity.

In Chapter 1, we discussed that the hegemony of Western discourses in
disaster studies and their application to disaster risk reduction cannot be
dissociated from their colonial heritage as it has been made possible only
because Western scholars and Western institutions have had the opportunity

to settle and conduct research in Latin America, Africa, Asia and the Pacific in the first place. This is a privilege seldom offered to non-Western scholars and institutions. As such, this imaginary representation of the world does not differ from what Said (1978) called Orientalism or the study of the Orient by Western historians, geographers and anthropologists over the past centuries. In Said's (1978, p. 67) own words:

> The Orientalist stage, as I have been calling it, becomes a system of moral and epistemological rigor. As a discipline representing institutionalized Western knowledge of the Orient. Orientalism thus comes to exert a three-way force, on the Orient, on the Orientalist, and on the Western "consumer" of Orientalism. ... Truth, in short, becomes a function of learned judgment, of the material itself, which in time seems to owe even its existence to the Orientalist.

Power and knowledge are intrinsically linked. Establishing a body of knowledge is a prerequisite to the exercise of power as much knowledge reflects the relations of power. It is through such differences between the West and the rest of the world, and their affirmation as absolute truth, that colonial policies and the contemporary disaster risk reduction agenda are imposed around the world. It is through the supposed backwardness of the rest of the world and its inability to deal with disasters that the West sustains its imperialist agenda. Thus, the body of skewed knowledge of what we call disaster in the non-Western world has legitimated decades of international disaster policy which has built on the transfer of experience and resources from the West to the rest of the world on the basis of the assumption that because the West suffers fewer casualties, we know best what works regardless of local contexts (Chapter 5). Bankoff (2001, p. 28) adds:

> In the scientific viewpoint, the West discovered a language of knowledge that has helped maintain its influence and power over other societies and their resources. In fact, natural disasters form part of a wider historical discourse about imperialism, dominance and hegemony through which the West has been able to exert its ascendancy over most peoples and regions of the globe. But the debate is not confined simply to geographies, however loosely defined; it is also a struggle over minds and, as such, has withstood the post-war dismantling of extensive colonial structures.

As such, intentionally or not, disaster studies has fuelled an imperialist disaster risk reduction agenda, which participates in the broader sustainable development agenda of the West as described by Escobar (1991, p. 329):

> This regime of representation assumes that it is up to the benevolent hand of the West to save the earth; it is the fathers of the World Bank, mediated by Gro Harlem Bruntland, the matriarch-scientist and the

few cosmopolitan Third Worlders who made it to the World Commission, who will reconcile 'humankind' with 'nature'. It is still the Western scientist that speaks for the earth.

Western scholarship therefore continues to dominate disaster studies and exert its influence over the rest of the world. It still constitutes the centre from which ideas emerge and eventually spread, maintaining centuries of combined hegemony and diffusionism (Said, 1978; JanMohammed, 1985; Blaut, 1993). The West is where research ideas shape up and where funding and equipment are available, where many researchers who study disasters in the rest of the world come from and where those who lead publications are affiliated. As Lyotard (1979, p. 74) once said, '*les jeux du langage scientifique vont devenir des jeux de riches, où le plus riche a le plus de chances d'avoir raison. Une équation se dessine entre richesse, efficience, vérité*[4]. Insidiously, the hegemony of Western scholarship further trickles down to within countries in the non-Western world where universities and scholars located in the capitals or other dominant cities exert the same power/control over institutions and researchers at the periphery, thus reflecting a progressive Westernisation of academia (Altbach, 2004; Mbembe, 2016).

The instruments of the hegemony of Western scholarship

The hegemony of Western scholarship has been possible only because of a coincidence of interests between Western scholars and their non-Western counterparts, the drivers and the partners, the principal investigators and the co-investigators in the lingua of contemporary research. In this way, active consent is initially suggested, in Gramsci's terms, on the side of the latter. Meanwhile, the former have given up some small chunks of funding and, to some extent, consider the ideas of the latter while legitimating the relevance of their research in a fashion, one may argue, akin to Memmi's (1957) description of the colonisers' constant desire to justify their action. Gramsci (1929–35, Q13, XXX, §18) elaborates on the nature of this process:

> Il fatto dell'egemonia presuppone indubbiamente che sia tenuto conto degli interessi e delle tendenze dei gruppi sui quali l'egemonia verrà esercitata, che si formi un certo equilibrio di compromesso (...), ma è anche indubbio che tali sacrifizi e tale compromesso non possono riguardare l'essenziale, poiché se l'egemonia è etico-politica, non può non essere anche economica, non può non avere il suo fondamento nella funzione decisiva che il gruppo dirigente esercita nel nucleo decisivo dell'attività economica[5].

Researchers based in non-Western countries in turn have seen such resources as an opportunity to gain research experience, build collaborations, access expensive equipment, conduct more field work and publish their findings in international journals, thus boosting their own careers. However, decision-making as well as intellectual and financial leadership most often remains within the hands of Western scholars. This skewed convergence of interests has created an equilibrium that underpins the hegemony of the West in disaster studies, favouring a few local scholars who constitute a form of local elite and whose role in supporting hegemonic agenda is crucial (Guha, 1982). As Spivak (1987, p. 145) once said:

> the putative center welcomes selective inhabitants of the margin in order better to exclude the margin. And it is the center that offers the official explanation; or, the center is defined and reproduced by the explanation that it can express.

Indeed, the co-investigators or partners of the Western scholars who somehow benefit from the system are but the tip of the iceberg. What about the multiple research assistants, interpreters and students who collaborate on projects led by Western scholars in contexts with which they are unfamiliar? That is, those whom Turner (2010) calls the ghost-workers or the subalterns of disaster studies? Some of them may be included down the list of co-authors of publications. Some are not. Some students may benefit from bursaries to pursue their postgraduate studies in Western universities but is this not a perpetuation of the Western hegemony and diffusionism when non-Western scholars learn how to research the Western way (Salazar, 1991; Altbach, 2004)? Was Foucault right when he suggested, during a famous television debate with Chomsky in 1971, that 'the university, and in a general way, all teaching systems, which appear simply to disseminate knowledge, are made to maintain a certain social class in power; and to exclude the instruments of power of another social class' (Chomsky and Foucault, 2006, p. 40)?

Such 'humanitarian' gestures and collaborations with local scholars, made through a clear relationship of domination, are essential to understanding the solidity and durability of the Western hegemony in disaster studies and its extension to disaster risk reduction. As Chatterjee (1983, p. 59) suggests, 'domination must exist within a relation. The dominant groups, in their exercise of domination, do not consume and destroy the dominated classes, for then there would be no relation of power, and hence no domination'.

The hegemony of Western scholarship in disaster studies is further accentuated by the continuing expansion and growing influence of the publication industry, controlled by Western publishers who are increasingly driven by indicators such as the number of citations the articles they publish receive. As such, only articles that match standardised expectations of how an argument unfolds and how an article is structured are deemed worth publishing. Thus, other world views/senses and approaches to research are overlooked (Canagarajah, 2002). In disaster studies, an article written in English with a title which refers to vulnerability is therefore much more 'marketable' than others written in, say, Nahuatl, Wolof or Telugu, all of which must use long sentences to capture something that only approximates to what Westerners usually mean by vulnerability. Within such an environment, is there any room to take on the challenge of the pioneers of the vulnerability paradigm to emphasise local and indigenous ontologies and epistemologies in understanding disasters?

Scholars of disaster studies are also encouraged to publish as quickly as possible following major disasters. As such, the 'gold rush attitude' is undoubtedly the product of the realms of the contemporary scientific scene where researchers need to produce a large number of publications every year in scientific journals and are being 'measured' using h-index, i10-index, Field-Weighted Citation Impact, and so on. This behaviour is first encouraged by the journals themselves as the first publications will be cited by the next wave of publications. Being the first to conduct research in the aftermath of a disaster thus provides a 'competitive advantage', in the words of Stallings (2007, p. 61), on the 'publication market'. Publishing as much and as soon as possible after a disaster is also encouraged by the researchers' own institutions, as the volume of publications increasingly becomes a guarantee of 'excellence', and is often a requirement for a continuing employment for neoliberal campuses, as Cupples (2011) pointed out. The above-mentioned chain of pressure is even longer, as universities in turn are attracted by the carrot of the world universities' rankings. Forty years ago, Hewitt (1983, p. 3) ranted:

> Contemporary natural disasters research is certainly rich in the results of scientific inquiries, whether in geophysics or the psychology of stress. The applications of scientific research are not, however, its definitive feature. It may have internal coherence or at least conviction. That does not alter my sense that it capitalises rather arbitrarily upon scientific discovery. Indeed it accords with "the facts" only insofar as they can be made to fit the assumptions, developments and social predicaments of dominant institutions and research that has grown up serving them. Moreover, my assessment of it leads me to believe that such developments have become the single greatest impediment to improvement in both understanding of natural calamities and the strategies to alleviate them.

The hegemony of Western scholarship is furthered by the continuing expansion of open-access journals that charge (very) expensive publication fees. Isn't it the ultimate form of imperialism when only the wealthy can publish and circulate their knowledge among the less affluent while contributing to the prosperity of Western publishers along the way? The traditional approach to publishing, that involves paywalls, is not much better but at least it offers an opportunity to everyone to publish while any researchers can approach any author to seek a copy of their publication.

The rapid expansion of these expensive open-access journals reflects the increasing pressure to publish that all scholars feel, both in the West and elsewhere (Danell, 2011). This is evidence of increasing accountability towards centres to the detriment of the peripheries in all their diversity. As Lyotard (1979, p. 76) once said, '*dans le discours des bailleurs de fonds aujourd'hui, le seul enjeu crédible, c'est la puissance. On n'achète pas des savants, des techniciens et des appareils pour savoir la vérité, mais pour accroître la puissance*'[6]. This statement reflects the increasing power of neoliberal academic institutions and associated funding agencies over researchers who are often struggling to secure their own position within the system and thus, consciously or unconsciously, are forced to passively consent to perpetuating the hegemony (Altbach, 2004; Mbembe, 2016). As Chakrabarty (2000, p. 23) brilliantly summarises, 'the globality of academia is not independent of the globality that the European modern has created'.

On conduct in a growing field of scholarship

Disaster studies has experienced a very fast expansion since its inception as a formal field of scholarship with Prince's (1920) PhD thesis more than a century ago. The quickly increasing and vast amounts of contemporary publications on what we call disaster mirror this momentum (Figure 3.3).

There are multiple reasons for this growing interest in studying disasters. The large and high-profile events we discussed earlier in this chapter, widely covered by the media, have stirred a genuine commitment to help alleviate hardship at both local and international levels. For instance, the scope and impact of the 2004 tsunami in South and Southeast Asia fostered significant interest among international researchers and have, as such, boosted scholarship on disaster worldwide. At the national level, disaster research in the Philippines exploded following typhoons Ondoy and Pepeng, which struck Manila in the span of a few days in 2009. Similarly, the 2010 earthquake has generated an increasing scholarly interest in disasters in Chile. Growing interest in studying disaster also results from increasing funding opportunities offered by national governments, international organisations and non-governmental organisations. As we see in Chapter 5, these organisations have to comply with the injunctions of international agreements, such as the Sendai Framework for Disaster Risk Reduction. As such, they need a clear picture of the issues they deal with as well as 'evidence' of the impact of the actions they take to tackle these issues.

This growth in scholarship has led to the creation of numerous networks and teams of researchers as well as specific institutions dedicated to studying what we call disaster. A profusion of new opportunities have also emerged to share research outputs. These include more than 80 academic journals publishing in English and focusing exclusively on the multiple dimensions of disasters across the hazard–vulnerability binary as well as an abundance of specialised books (Alexander et al., 2020). In addition, there are currently multiple conferences, seminars, workshops and other

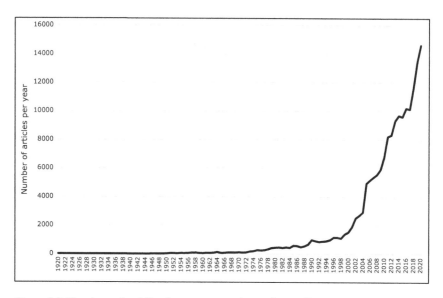

Figure 3.3 Number of publications per year that refer to disaster in their title, keywords or abstract between 1920 and 2020 (data from Scopus)

meetings of scholars and, in a growing number of instances, policy-makers and practitioners. In fact, it has become virtually impossible to keep up-to-date with all the current publications and events in the field of scholarship that is disaster studies.

The traction that disaster studies has gained over the past 20 years has obviously accentuated the issues we discussed earlier in this chapter with regard to unequal power relations among researchers. It has also reinforced the hegemony of Western ontologies and epistemologies in understanding what we call disaster in the West. To address some of these issues, we have recently called for a 'code of conduct' to be devised among scholars of disaster (Gaillard and Peek, 2019). This call draws upon three principles: first, that scholars have a clear purpose for their research; second, that they contribute to addressing a legitimate gap in knowledge informed by local voices; and, third, that research be conducted ethically, show respect to local values and benefit those who are affected and also that scholars foster a genuine and fair collaboration between local and outside researchers.

This call for attention to the issues we discussed in this chapter has gained significant attention (e.g. Alburo-Cañete, 2021; Aragón-Duran et al., 2020; Chmutina et al., 2020; Oulahen et al., 2020). However, it has also been misinterpreted in some instances (Kendra and Wachtendorf, 2020). Our intention is not to suggest a strict and rigid 'code'. That would be antithetical to our argument in Chapter 5 of this book. Rather, we suggest a flexible set of questions that encourage researchers to build upon local social and cultural values and draw upon the ability of local scholars and those affected by disaster to lead research as much as possible. As such, we do not propose that there be a one-size-fits-all series of guidelines to be rolled out everywhere in a blanket fashion. In fact, our intention is for this set of questions to be devised from the bottom up, through local and regional meetings, before being elevated to and discussed at an international level.

Neither is our intention for this set of principles to be binding or embedded within the institutional structure of any international organisation, such as those of the United Nations Office for Disaster Risk Reduction. Rather, we suggest that the devising of our proposed set of principles be led by local researchers, in all their diversity. We understand that there will always be issues around representation, but we suggest that as many of these researchers as possible come up together in a neutral forum that could be, we propose, the Global Platform for Disaster Risk Reduction. Our intention is that the set of questions we suggest not be co-opted or appropriated by any single research or academic institution or network.

There have already been some discussions in this direction but most have been confined to specific disciplines, such as psychology and biomedical sciences (e.g. Collogan et al., 2004; Sumathipala et al., 2007; O'Mathúna, 2010), bound to some few countries such as the Philippines and Indonesia (Philippine Health Research Ethics Board, 2011; the Ministry of Research, Technology and Higher Education, 2017) or limited to relatively narrow networks of scholars (https://www.ipetitions.com/petition/power-pres-tige-forgotten-values-a-disaster). Given the multidisciplinary and international nature of disaster studies, it seems important to broaden the scope of these discussions, across disciplines and countries, for research in time of disaster as well as for scholarship grounded in the everyday lives of people who may be affected by what we call disaster.

Ultimately, our concern is that researchers, whether local or international, conduct fair, just and grounded research – research that fosters the leadership of local or grounded researchers (or both) and respects local values and priorities, research that favours local ontologies and epistemologies over the sole Western theories, concepts and methods, which, we recognise, are still likely to be valid and relevant within their cradle. Such an approach, in fact, is aligned with the original tenets of the so-called vulnerability paradigm, and upholding its promises requires us to explore postcolonial opportunities to researching disasters. An agenda that we outline in conclusion of this book.

Notes

1 We recognise and discuss the limitations of this and other databases in Chapters 5 and 10.
2 '*The fire of 1631 produced, if I may, a typographic epidemic with both knowledgeable and ignorant people rushing in proportions that we cannot describe to commemorate the terrible explosion*'. Our translation.

3 The ideas in this paragraph owe much to the sharp and salient mind of Christopher Gomez.

4 *The games of scientific language become the games of the rich, in which whoever is wealthiest has the best chance of being right. An equation between wealth, efficiency, and truth is thus established*. From the English edition translated by Geoff Bennington and Brian Massumi and published by the University of Minnesota Press in 1984.

5 '*Undoubtedly the fact of hegemony presupposes that account be taken of the interests and the tendencies of the groups over which hegemony is to be exercised, and that a certain compromise equilibrium should be formed (…). But there is also no doubt that such sacrifices and such a compromise cannot touch the essential; for though hegemony is ethical-political, it must also be economic, must necessarily be based on the decisive function exercised by the leading group in the decisive nucleus of economic activity*'. From the English edition translated by Quintin Hoare and Geoffrey Nowell Smith and published by International Publishers in 1971.

6 '*In the discourse of today's financial backers of research, the only credible goal is power. Scientists, technicians, and instruments are purchased not to find truth, but to augment power*'. From the English edition translated by Geoff Bennington and Brian Massumi and published by University of Minnesota Press in 1984.

4 The quest for pantometry

The project of modernity that underpins contemporary understandings and discourses on disaster has been sustained by a quest for truth: one and only truth informed by reason and the precepts of the Enlightenment. Truth which entails that disasters sit at the interface between nature/hazard and culture/vulnerability.

Such a pursuit of truth requires us to provide proof or evidence of any claims. Any scientific discovery or narrative needs to be legitimate in the eyes of both scientists and those who will make use of this knowledge, whether they are policy-makers, practitioners or industry. Legitimation, Lyotard (1979) says, is crucial to the project of modernity and the associated advancement of science since it is these proofs that make scientific knowledge irrefutable. This quest for truth has turned into a 'worship of "objectivity"' (Constantino, 1978, p. 283).

The project of modernity's quest for one single truth has long required that the world be perceived through a homogenous lens, so that things, subjects, individuals, processes, phenomena, and so on can be reduced, measured and compared. In such a positivist line of thinking, numbers and quantifications have provided the most reliable and irrefutable form of evidence and legitimation (Horkheimer and Adorno, 1947). As a result, pantometry, or the quest for universal measurement, supported by the development of sciences, especially mathematics, has been an essential dimension of science and policy. Crosby (1997) actually argues that it is Europe's increasing command of measurements of all sorts, and of mathematical techniques in particular, that allowed for its expansion and imperial project, which would flourish in the eighteenth century and eventually support the project of modernity.

The prominence of quantification in the search for irrefutable evidence has never vanished, in spite of the emergence of other diverse forms of epistemologies, including those associated with constructivism, that argue that there is no such thing as a single view of the world and that people and society cannot be reduced to numbers. Compared with numbers and measurements, alternative forms of evidence provided by words and writing, through ethnographic studies, and visuals in participatory research have received only marginal attention in informing policy and practice for reducing the risk of disaster.

DOI: 10.4324/9781315752167-4

In disaster studies, this quest for evidence has informed the study of natural hazards and also that of the vulnerability of people and society. The robustness of the evidence is to inform relevant, rational, efficient and strategic decision-making in policy and actions towards disaster risk reduction:

> (...) demand is growing for a more prominent and effective role for science and technology in providing evidence for policy, with the international community recognising that successful disaster risk reduction (DRR) depends on it. As such, science is to be included as a core aspect of the Post-2015 DRR Framework.
>
> (Carabine, 2015)

> The new framework [i.e. the Sendai Framework for Disaster Risk Reduction] should enable critical public policies that are informed by evidence from science and tools from technology to address disaster risk in publically owned, managed or regulated organizations and infrastructures, as well as support actions by households, communities, businesses and individuals.
>
> (Aitsi-Selmi et al., 2015)

The link between science and policy/action towards disaster risk reduction is therefore one of legitimation and guidance. The obvious question is whose evidence informs whose decision and in whose interests?

This question cannot be fully answered without considering the underpinning ontologies that sustain our understandings of disaster as a concept but also as an object of research and focus for policy and action. To meaningfully address the question therefore requires that we locate the inquiry within the binary assumption that disasters sit at the interface between nature/hazards and culture/vulnerability. This chapter particularly tackles the kinds of evidence that inform the study of disaster and, by extension, disaster risk reduction.

We focus this discussion on the specific concept of resilience. Over the past couple of decades, the concept of resilience has gained growing attention in academic and policy discourses as well as within circles of practitioners (e.g. Alexander, 2013; Manyena, 2006; Weichselgartner and Kelman, 2015). The concept has been picked up by researchers, international organisations, non-governmental organisations (NGOs), funding agencies and industry, so that it has become one of the main buzzwords in the fields of development and disaster risk reduction. Notwithstanding the loose conceptual framing that we discussed in Chapter 2 of this book, strengthening resilience has become the priority of most (if not all) agendas for disaster risk reduction at all scales, from the Sendai Framework for Disaster Risk Reduction and international funding agencies' priorities to national policies and the everyday initiatives of practitioners.

As a result, international organisations, governmental agencies and NGOs have progressively felt the need for measuring resilience in order to (1) prioritise policies and actions, (2) monitor progress and (3) foster accountability (Béné, 2013; Levine, 2014). This demand for evidence and measurement has stimulated an incredible flurry of approaches, methods and tools designed by researchers, as well as international organisations and NGOs, to capture the multiple dimensions of resilience at a variety of scales.

The following sections provide a non-exhaustive but hopefully representative and critical overview of these approaches, methods and tools, framed within the different paradigms that have guided disaster studies and disaster risk reduction over the past century. Notably, though, the subsequent review and discussion purposefully exclude the specific and very large field of resilience measurement in psychology for which multiple reviews are already available (e.g. Norris et al., 2008; Windle et al., 2011).

What kind of evidence for what kind of measurement?

We have seen in Chapter 2 how the concept of resilience is currently used in ways that do not always reflect its original objective to build on people's intrinsic ability to deal with natural hazards and disasters. As a result, the diverse approaches for measuring resilience to natural and other hazards reflect different ethos that are grounded in divergent and often opposing epistemologies. These ultimately mirror the different paradigms that have emerged in disaster studies and how the latter have influenced policies and actions for reducing disaster risk (Chapter 2). The three streams of methods identified by Chambers (2007a) in the context of poverty assessment provide a useful framework for comprehending the different ways that resilience is approached. These three streams are (1) economic reductionism, (2) anthropological particularism and (3) participatory pluralism.

Economic reductionism and the hazard paradigm

Economic reductionism refers to quantitative and non-contextual methods grounded in a positivist view of the world, including of disaster risk. In line with the tenets of the hazard paradigm, it emphasises the importance of *extra*-ordinary natural hazards that overcome the ability of those affected to cope.

Methods to assess resilience are, as a result, quantitative and lead to computing scores, ranks and indexes. These reductionist methods are driven by outside experts who design questionnaires and other extractive tools based on their generalised assumptions of what resilience is and what it means. Local people who face natural hazards and disasters are thereby passive respondents of surveys and other censuses. Surveys to assess resilience are conducted in different regions of the world following similar approaches, so that the outcomes are comparable (White, 1974). These allow for the design of universal theoretical frameworks such as the choice tree of adjustment to natural hazards, which has been applied from individual to societal levels (Burton et al., 1978).

This reductionist approach to assessing both hazards and resilience has underpinned top-down transfers of knowledge, technology and experience from the wealthiest and most powerful regions of the world, which

are supposed to be safer because of their larger technological and economic resources, towards the poorest and less powerful ones, which are
considered more vulnerable and unable to cope alone (Bankoff, 2001).
Disaster risk reduction therefore has long been predominantly a technocratic process based on the judgement of experts, who most often do not
face hazards and disasters themselves. It entails command-and-control initiatives meant to be for the good of those at risk who are deemed unable to
make informed judgements because they lack a command of the numbers
at stake. These initiatives include building infrastructure to prevent hazards and an arsenal of actions geared towards better controlling those at
risk by providing them with scientific information to raise their awareness
of natural hazards and 'managing' their response to such events should
they occur (Chapter 5).

Anthropological particularism and the vulnerability paradigm

Anthropological particularism draws on qualitative methods geared toward
providing contextual and rich descriptions of local realities. These have
long been used in disaster studies to examine the response of people to disaster and more recently to assess their resilience to natural hazards. Resilience to natural hazards is here viewed in a dialectical relationship with the
concept of vulnerability.

This approach to resilience in dialogue with vulnerability grounds it
within the context of the so-called vulnerability paradigm that we discussed
in Chapter 2. However, it does not mean to reflect the entirety of the vulnerability paradigm as framed and summarised by Wisner et al. (2004).
Nor does it mirror a desire to measure in mathematical terms. Anthropologists and geographers, in particular, have focused on providing multiple and detailed qualitative accounts of people's resilience which have
unpacked, or challenged, the importance of natural hazards in explaining
the occurrence of disasters and their impact (Hewitt, 1983; O'Keefe et al.,
1976; Oliver-Smith, 1996; Torry, 1979b).

Understanding people's resilience to natural hazards in relationship to
their vulnerability requires fine-grained studies relying upon qualitative
and often ethnographic research methods to better elicit the unique realities of everyday life. These include tools such as semi-structured interviews,
life stories, participant observations, and focus groups designed to foster
interaction between outside researchers/practitioners and local people.
The outcomes of such studies are qualitative and hardly comparable from
one place to another.

Anthropological particularism and the vulnerability paradigm have led to
challenging the dominant technocratic approach to disaster risk reduction.
They have gained much ground over the past two decades, although their
direct translation into policy and practice has been relatively limited. In this
perspective, enhancing people's resilience can result only from addressing
the root causes of people's vulnerability. Because these are largely exogenous

to those at risk, strengthening resilience requires the intervention of those with power in order to grant access to resources to those vulnerable and, as a result, strengthen their livelihoods. This depends upon a radical shift in current policy and practice that warrants strong political commitment (Wisner et al., 2012).

Participatory pluralism and the capacities spin-off approach

Participatory pluralism refers to approaches geared towards fostering the participation of people at risk, in all of their diversity, in assessing their own resilience. It draws on the assumption that those at risk display capacities/resilience in facing natural and other hazards. As such, it supports the capacities/resilience spin-off approach to the vulnerability paradigm that emerged in the late 1980s.

Participatory approaches for assessing resilience to natural hazards have gained significant traction among practitioners. These are currently gathered under the umbrella of Participatory Learning and Action. Participatory Learning and Action refers to methods and attitudes designed to empower those at risk to share, analyse and enhance their knowledge of disaster risk as well as to plan, implement, monitor, assess and reflect in their efforts towards disaster risk reduction (Chambers, 2007b). The emphasis on attitudes and behaviour is particularly important here. Indeed, participatory approaches to assessing resilience often put forward unique sets of tools to produce visual data intelligible to all, including to those who may not be able to read, count and write. If these tools are important to elicit people's knowledge, they need to be properly facilitated in order to hand over the stick and shift power relations for the benefit of those at risk. To put the last first, to echo Chambers's (1983) famous saying: the process through which resilience is assessed is ultimately more important than the sole outcomes of the evaluation.

Participatory pluralism therefore suggests that assessing and enhancing resilience are the prime responsibilities of those at risk, who will take the lead in designing the most appropriate disaster risk reduction strategies in the context of their everyday and longer-term priorities. Outside stakeholders, including scientists and government agencies, are to provide only external support to sustain people's initiatives and foster the transfer of experiences across spatial scales (Delica-Willison and Gaillard, 2012). These initiatives often are garnered under the framework of 'community-based/led/managed' disaster risk reduction despite the challenging nature of the concept of 'community' (Chapter 7).

Measuring resilience in practice: a brief review of methods and tools

The past couple of decades have witnessed the emergence of a flurry of methods and tools to provide evidence and assess resilience in a variety of contexts. These have focused on multiple scales, from household, organisation and 'community' (whose meaning, once again, is highly contested)

Table 4.1 Characteristics of the three main types of approaches to measuring resilience to natural hazards and disasters

Approach	Economic reductionism	Anthropological particularism	Participatory pluralism
Ethos and principle	Resilience as an attribute / a reflection of losses	Resilience as a process	Resilience in the eyes of those facing natural hazards and disasters
Methods	Quantitative	Qualitative	Participatory
Role of outsiders	Data collector	Participant observer	Facilitator
Role of local people	Respondent	Social actor	Analyst
Mode	Extractive	Interactive	Self-organising
Contribution to knowledge	Comparable numbers and indexes	Social and cultural insights	'Surprises'
Outputs	Tables, graphs, maps	Rich descriptions	Tables, diagrams and charts, maps

(adapted from Chambers, 2007a)

to city/province/region and country levels. This section does not attempt to provide an exhaustive review of all these methods and tools. Rather, it endeavours to illustrate how the three main streams of measurement, discussed in the foregoing section, materialise in practice (Table 4.1). For those readers interested, more detailed and recent reviews of existing methods and tools are available in both the academic (e.g. Beccari, 2016; Ostadtaghizadeh et al., 2015) and policy (e.g. Levine, 2014; Winderl, 2014) literature.

Economic reductionism: indexes and other quantitative measurements of resilience

A large and diverse set of tools for the quantitative measurement of resilience has been developed (e.g. Beccari, 2016; Levine, 2014; Ostadtaghizadeh et al., 2015; Winderl, 2014). These tools draw on two fundamentally different assumptions, independent of how resilience is actually defined. On one hand, resilience is considered through the lens of past and potential losses and thus mirrors the impact of disasters. On the other hand, resilience is viewed as an attribute of people or places. These two different approaches for apprehending resilience have led to two parallel streams of quantitative assessments. For both streams, data is collected, manipulated and analysed by outsiders (e.g. researchers, government officials and NGO staff). Whenever interactions with local people occur, they are the passive respondents of an extractive process (Table 4.1).

Assessing resilience on the basis of losses requires an appropriate and reliable dataset, whatever the scale of the analysis. Data on losses is eventually integrated with a range of social and economic indicators through various equations. Indicators are identified through questionnaire surveys and/or secondary data from census organisations, depending on the scale of analysis. The attempt of Hallegatte et al. (2016) to measure the resilience of Mumbai, and eventually a number of countries other than India, to flooding constitutes a good example of one such approach. The analysis relies upon an econometric estimation and modelling of consumption and output losses for the whole city of Mumbai and then of its inhabitants' asset and income/welfare losses in correlation with their social and economic status as assessed through various indicators. The resulting model was eventually calibrated with household surveys conducted in the city and scaled up to 90 countries. For the latter, resilience was calculated as the ratio between expected asset losses and unexpected asset losses and, as a result, expected welfare losses, in the context of exposure to flood hazards and the latter's return period. Ultimately, resilience is compared with a country's gross domestic product.

Measurements of resilience as an attribute of people or places are conducted at both country and sub-administrative levels, be these neighbourhoods, cities or provinces, and rely on the available secondary quantitative indicators or questionnaire surveys (or both). Diverse indicators are used as proxies for the multiple dimensions of resilience and are compiled into composite indexes. A large number of variables from census and survey data are usually normalised/standardised and then scaled/weighted and aggregated using various equations. For example, Manyunga's (2009) Community Disaster Resilience Index draws on 75 variables, eventually aggregated into 15 sub-indexes and then into the main resilience index at the scale of counties and parishes in the United States of America. Cutter et al. (2010) follow the same process at the same scale using 36 variables compiled into five sub-indexes to form a main Disaster Resilience Index. For both indexes, the results and evidence ultimately are presented in maps of the different counties to visualise their different levels of resilience and to facilitate decision-making and the prioritisation of actions towards strengthening disaster risk reduction.

Other composite resilience indexes designed to reflect resilience as an attribute combine secondary data with primary information to capture more specific dimensions of resilience and tailor measurements to local contexts. Primary data is usually collected through questionnaire surveys, the outcomes of which are translated into numbers. This quantitative data is gathered to capture multiple dimensions of resilience through selected indicators and proxies. For example, the Climate and Disaster Resilience Index (CDRI), designed to assess the resilience of Indian cities, considers 125 indicators divided across five categories, each of which includes a specific set of multiple variables (Shaw et al., 2010). Data is collected through a couple of questionnaires to be filled out by government officials at regional and then local levels. The weighted mean of the responses for each category is calculated through a formula and then indicators are combined and graded along a scale of four degrees of resilience. As for the previous set of indexes, the resulting indexes and evidence are presented as maps of the cities to visualise the different degrees of resilience of each neighbourhood.

Finally, there are resilience indexes that rely exclusively (or almost exclusively) on primary data to better reflect the views of local stakeholders. Data usually is collected through interviews, focus groups or (more frequently) questionnaire surveys to provide quantitative evidence that is standardised and collated through various equations. The method of Béné et al. (2016) of measuring resilience in Fiji, Vietnam, Sri Lanka and Ghana provides an example of such an approach. Gender-disaggregated focus groups initially are conducted with people of different occupations to elicit a set of qualitative data that is used to design a couple of questionnaires focusing on resilience per se and quality of life. The questionnaires provide a set of 34 variables that are aggregated into a model to compute an index of resilience. The results and evidence are presented through tables and graphs crafted to provide explanations for different patterns of resilience.

Anthropological particularism: qualitative measurement of resilience

All sorts of qualitative studies are geared towards assessing the resilience of people and places. Most of these emerge from the social sciences and try to explore the deep-seated mechanisms that underpin the process of resilience at different scales. Some are standalone studies for the sake of academic research, whereas others are designed to reflect upon the outcomes of a particular project. Finally, some are geared to inform policy and practice. The myriad of available studies can be roughly classified into three streams of approaches reflecting how much filtering/analysis is carried out by the researcher. In all three streams, however, the researcher is a participant-observer who engages in an interactive relationship with local people who are social actors (Table 4.1).

The first of the three streams consists of the raw accounts and stories of people at risk or from those affected by disasters. These testimonies usually are collected through extractive tools such as interviews which reflect researchers' expectations. However, they are meant to provide vivid evidence of how people feel about resilience. These accounts are highly contextual and can hardly be compared with each other. Despite this, they are often compiled into books or reports to show the diversity of people's experiences and their needs when facing natural hazards and disasters. Ride and Bretherton's (2011) compilation of testimonies from five countries – Indonesia, Pakistan, Solomon Islands, Kenya and Myanmar – illustrates this approach well. It draws on a series of academic case studies involving semi-structured interviews conducted in local languages with an average of 11 informants in each country. People's stories of disasters and resilience then are reported in a textual format relying on extensive quotations.

The second array of studies entails further involvement on the side of the researcher who often draws on a more diverse range of ethnographic tools, including interviews/life stories and observations, to collect qualitative evidence of people's resilience in facing natural hazards and disasters. The analysis of this data involves a higher degree of data manipulation though a wide range of codified methods that includes, for example, content or thematic analysis. Computer programmes are also increasingly used to tease nodes and themes out of the datasets. Hastrup's (2011) and Simpson's (2014) accounts of how people overcame the impact of two disasters in India – the 2004 tsunami in Tamil Nadu and the 2001 earthquake in Gujarat, respectively – are excellent examples of this type of assessment of resilience. These studies rely on years of field research that provides temporal grounding. They are narrated in text with particular attention to certain details in order to provide a fine-grained analysis of the drivers of people's resilience. Such evidence of resilience is also highly contextual and constitutes standalone studies that hardly can be directly compared.

A third level of qualitative assessment of resilience aims at providing proactive frameworks to anticipate and measure resilience in disaster risk reduction projects. These frameworks emphasise key and usually broad components of resilience that need to be considered in designing projects and measuring progress. These frameworks usually do not refer to particular tools, nor do they require any quantitative measurements of any indicators. For example, Buckle (2006) identifies seven factors that support people's resilience in facing natural hazards and disasters. These are knowledge of hazards, shared community values, established social infrastructure, positive social and economic trends, partnerships, communities of interest, and resources and skills. These factors are provided to guide and prioritise practitioners' initiatives towards strengthening resilience.

Towards participatory pluralism: toolkits and characteristics of resilience

Another set of tools and methods for assessing resilience has emerged and proliferated over the past two decades, encouraging the participation of more people in the measurement of what resilience means for them. In fact, it is likely that these days each and every NGO active in the field of disaster risk reduction has its own toolkit for assessing and measuring resilience.

Notably, though, none of these tools and methods is likely to be fully participatory as they (1) all draw on predefined assumptions and characteristics of what resilience encompasses and (2) involve some level of data manipulation by outsiders, often by the staff of NGOs. However, their gathering into relatively loose and flexible toolkits, often referred to as 'participatory tools', provides enough space for local stakeholders to make them fit specific and unique needs. In fact, Twigg (2009: 11) suggests that predefined characteristics of resilience should be customised and modified. The role of outside stakeholders is therefore crucial and should be that of a facilitator supporting local people in collecting and analysing their own data (Table 4.1).

Twigg's (2009) *Characteristics of a Disaster-Resilient Community* is probably the most popular approach to measuring resilience that gives enough space for people at risk to express their own views of what resilience means for them. These characteristics cover five thematic areas broken down into 28 components of resilience, which are subdivided into 161 characteristics of resilience. These serve only as points of reference or 'signposts' to assist practitioners in identifying context-specific evidence and tangible indicators to measure resilience. These characteristics do not suppose the use of any specific tools but rather encourage practitioners to rely upon existing and flexible toolkits such as the Hazard, Vulnerability and Capacity Analysis (HVCA) approach. In that sense, a genuine participatory process may lead to indicators that can reflect the *Characteristics of a Disaster-Resilient Community.*

Another significant example of these toolkits is CoBRA (Community-Based Resilience Analysis), developed by the United Nations Development Programme to assess resilience at the household level (United Nations Development Programme Drylands Development Centre, 2014). CoBRA draws on predefined characteristics and indicators of resilience designed after the Eurocentric/Western set of resources that make up the Sustainable Livelihood Framework (i.e. natural, physical, social, financial and human). However, data is collected through interviews with key informants and focus groups which make use of flexible tools, often labelled 'participatory', and thus provide local people with reasonable opportunities to express their views of what resilience means for them. Nonetheless, the data collected is then manipulated and analysed by external stakeholders to guide and prioritise their own activities.

Data collected and analysed through participatory toolkits is usually presented as graphs, diagrams, charts and maps that are accessible to all those who face natural hazards and disasters, including those who may struggle to read, count and write (Table 4.1). The visual dimension of these forms of measurement of resilience is supposed to be one of their main strengths as it allows cultural and literacy barriers to be overcome (Chambers, 2010).

Inherent strengths and limits of the method(s) and the search for compromise

All approaches to assessing resilience have their own strengths and shortcomings (Table 4.2). The reductionist approach to quantifying resilience is relatively quick to set up and provides tangible evidence that allows comparisons across places. They therefore facilitate decision-making and prioritisation in policy. Over time, policy-makers and practitioners have, in fact, become number-savvy, to the point that decision-making often has become a matter of juggling with figures and statistics. Numbers and figures further speak to donors and upper-government agencies that call for upward

Table 4.2 Main strengths and limitations of the three main types of approaches to measuring resilience to natural hazards and disasters

Approach	Strengths	Limitations
Economic reductionism	- Quick to set up - Comparable numbers	- High level of generalisation - Biased by outsiders' choices
Anthropological particularism	- Contextual details - Long-term processes	- Time consuming - Hardly comparable
Participatory pluralism	- Reflects people's own and diverse views - Addresses actual local needs	- Highly dependent on facilitators' skills - Hardly comparable

accountability from their funding beneficiaries. Yet quantitative evidence and measurements of resilience often fail to capture the reality of those at risk, whose experiences differ in time and from one household and one place to another. They are also often biased by choices made by outsiders that include focusing on easily accessible places at favourable times of the year. For this reason, they have been called 'quick-and-dirty' by Chambers (1984). Quantitative measurements also require large and varied datasets, which may not be available in the first place and whose reliability and validity may be questionable.

Still, reductionist approaches to assessing resilience are the most common these days, not only among scholars of disaster studies but also in circles of practitioners. This reflects the importance of upward accountability in the practice of disaster risk reduction. Indeed, many practitioners, be they the staff of NGOs or local government officials, often feel an obligation to report the tangible outcomes of their activities to donors and upper agencies. It is therefore appealing to many to quantify increased resilience as numbers and figures which often are believed to provide the absolute truth in opposition to qualitative evidence. This obligation for upward accountability is deeply entrenched in the obsession of international donors and national governments with accountancy that is associated with a wide range of conditions placed on the project: its time frame, the role of the different stakeholders, and so on (Chapter 7).

Qualitative methods for assessing resilience provide fine-grained views of people's realities and contribute to capturing less tangible aspects of their ability to face natural hazards and disasters. They prove particularly strong in exploring causalities, including for understanding why people are resilient, or not, through the complex interaction of drivers at different time and spatial scales. Yet these approaches inherently reflect the relations of power between the researchers and the informants and how the former interpret the stories of the latter. In a famous critique of ethnography,

Clifford (1986a, p. 7) pondered that 'even the best ethnographic texts – serious, true fiction – are systems, or economies of truth. Power and history work through them, in ways their authors cannot fully control. Ethnographic truths are thus inherently partial – committed and incomplete'. Furthermore, anthropological particularism has been criticised for being long-and-dirty, in the sense that it requires long studies that often are outdated when finalised or that remain on the shelves of academic libraries and hence are of little use for policy and practice (Chambers, 1984). They also prove difficult to reproduce in space and time, which constitutes another impediment to their inclusion in policy at national and international levels. Ultimately, the use of qualitative methods for assessing resilience has largely been limited to academic research and has had little impact on policy and practice.

In contrast, participatory pluralism has spread broadly within circles of practitioners. Participatory approaches draw on people's knowledge and skills and foster the participation of those at risk in both assessing and strengthening resilience. In theory, they further reflect the diversity of people's realities and emphasise downward accountability towards those at risk (Chapter 7). Participatory methods are easy to set up and flexible but require facilitation skills and experience to encourage genuine participation and the transfer of power. Often, though, the process is *facipulated* and skewed to serve the interests of outside stakeholders who need to justify the 'involvement' of locals in activities they have designed beforehand in a typical upward-accountability approach (Cooke and Kothari, 2001). They often rely extensively on taxonomic categories of resources, which are reductionist in nature and reflect Western world views/senses such as that the land and the sky fit under natural resources rather than social entities or that gender is constituted of women and men identities only (Chapter 8). Furthermore, participatory approaches to measuring resilience are frequently distrusted by policy-makers who struggle to make sense of such highly context-specific evidence. In addition, participatory assessments of resilience often are disconnected from formal (government and scientific) initiatives.

Ultimately, the foregoing review of approaches for assessing resilience depicts a fragmented landscape where the three different methodologies for assessing resilience operate in silos. Attention has recently been given to bridging that gap which entails overcoming deep-seated epistemological challenges. The most prominent of them, when it comes to assessing resilience, has been to identify tools that allow all stakeholders to collaborate in the same activities, around the same tables and at the same times. These tools should be trusted by all actors and make people's view of their own resilience tangible to outsiders. They should facilitate comparisons across people, households and places in order to prioritise and inform policy decisions. In line with the precepts of the vulnerability paradigm, these tools also need to be integrated into broader disaster risk assessment and reduction frameworks which consider both the root causes of vulnerability and people's capacities. This allows for the integration of a diverse set of actions, including those from the bottom up and those from the top down (Gaillard and Mercer, 2013).

In this perspective, quantitative participatory methods (QPMs) have been suggested to pull together quantitative and qualitative as well as participatory assessments of resilience. QPMs generate what is known as participatory numbers (Chambers, 2003, 2007b) or participatory statistics (Holland, 2013). Participatory numbers are 'quantitative research information produced by those at the forefront of everyday development struggles, i.e. the poor and marginalised who are usually excluded from mainstream research initiatives supposed to assist in lifting their wellbeing' (Gaillard et al., 2016a, p. 1000). QPMs aim to attribute a 'value' to the qualitative and often intangible dimensions of people's resilience through a participatory research process in which local people define their own indicators and then analyse and monitor them themselves. It is therefore expected that participatory numbers are more likely to reflect people's realities while providing some tangible evidence for comparison and scaling up to inform decision-making beyond the place where numbers are produced (Chambers, 2010). QPMs and participatory numbers therefore are meant to provide an opportunity for pulling together 'the best of all worlds', to expand an expression coined by Barahona and Levy (2007). As such, QPMs reflect the ultimate quest for pantometry.

On Western heritage and the quest for pantometry

It is no surprise that the foregoing epistemological and methodological debate on how to measure resilience mirrors the theoretical discussion that has informed disaster studies over the past century. It is firmly grounded in the ontological assumption that disasters sit at the interface between the nature/hazard and culture/vulnerability. In this perspective, the ways resilience has been measured are (especially for methods associated with economic reductionism and anthropological particularism) firmly grounded in a Western tradition of scholarship and its extension to policy and practice. One may, in fact, argue that the epistemological and methodological debate discussed in the previous section is just one more dimension of the Hegelian dialogue that has shaped the development of disaster studies over the past century.

Reductionist approaches to measuring resilience are directly inherited from the West's quest for quantification inherited from the Enlightenment. As such, reductionist approaches support the project of modernity. Crosby (1997, p. 17) writes:

> The West's distinctive intellectual accomplishment was to bring mathematics and measurement together and to hold them to the task of making sense of a sensorially perceivable reality, which Westerners, in a flying leap of faith, assumed was temporally and spatially uniform and therefore susceptible to such examination.

If qualitative and ethnographic methods seem to critically differ from reductionism in their ethos and objectives, they nonetheless similarly mirror a Western tradition of scholarship, one grounded in a long tradition of ethnographic research and the West's quest to rationalise, classify and report in writing on cultures where world views/senses and ways of knowing follow different precepts (Clifford, 1988; Marcus and Fischer, 1986). As such, it does not really differ from the purpose of reductionist and quantitative approaches to understanding the world and disasters. As Clifford (1986b, p. 99) says, 'in the West, the passage from oral to literate is a potent recurring story of power, corruption and loss'.

This quasi-obsession for reducing disasters to words and numbers and visualising them on paper is a common thread across both quantitative and qualitative assessments of resilience and a key legacy of Enlightenment thinking. As Crosby (1997, pp. 228–229) summarises:

> the new approach was simply this: reduce what you are trying to think about to the minimum required by its definition; visualize it on paper, or at least in your mind (…), and divide it, either in fact or in imagination, into equal quanta. Then you can measure it, that is count the quanta. Then you possess a quantitative representation of your subject that is, however simplified, even in its errors and omission, precise. You can think about it rigorously. You can manipulate it and experiment with it, as we do today with computer models. It possesses a sort of independence from you.

Further parallels can be drawn with Clifford's (1986b, p. 99) critique of ethnographic writings: 'the very activity of ethnographic writing – seen as inscription or textualization – enacts a redemptive Western allegory'. A critique eventually expanded to arts and artefacts collections in museums or by private individuals (Clifford, 1988), which we briefly tackle in Chapter 5. The importance of visualising resilience and disaster on paper expands to the significance of the very mnemonic with which we opened this book. It also appears in the crucial role that databases, maps and plans have taken in both understanding and addressing disaster risk which we discuss in Chapter 5. It has even penetrated participatory pluralism and become one of its core tenets (Chambers, 1994; McKee, 1994). In fact, participatory methods have been praised for offering a new form of literacy (Robinson-Pant, 1996) that Chambers (1994, p. 1263) justifies:

> Visual methods can also empower the weak and disadvantaged. Visual literacy is independent of alphabetical literacy, and appears to be near-universal. Visual diagramming is thus an equalizer, especially when it is done using the accessible and familiar medium of the ground.

Nonetheless, participatory pluralism remains a more radical departure from anthropological particularism than the capacities spin-off is from the vulnerability paradigm. From an epistemological and methodological perspective, it constitutes a clearer step towards the third phase of the Hegelian historical quest towards truth and reason; yet one that is dislocated from what McEntire (2001) and Hilhorst (2004) have put forward as the third paradigm in disaster studies. As such, it does not help firm up the existence of a clearly distinct and fully delineated third paradigm.

Rather, participatory pluralism, as an approach to understanding the world including to what the West means by disaster, from the eyes of people who do not necessarily share the same Eurocentric heritage, offers an opportunity for stepping away from the ontological assumption that disasters sit at the interface between nature and culture. In theory, participatory pluralism is thus the most promising pathway out of the hegemony of Western understandings and discourses on disasters. In practice, though, we have seen in the previous section that it is often skewed towards the expectations of outsiders and its process is frequently *facipulated* to meet these expectations.

The recent emergence of QPMs is interesting as it ultimately reflects an organic tendency to return to numbers and quantification as the best opportunity to rationalise and provide proofs and, in Lyotard's terms, to support both scientific understandings of disasters and actions to reduce disaster risk. There is therefore, in the ways we have been measuring resilience, an organic and continuing tendency to ultimately return to the ontological and epistemological belief that there is one truth with regard to resilience. This entails that there is one such thing as resilience to Nature's challenges and that it can be measured/assessed and represented in numbers, words, diagrams and maps in some ways. This observation is also true for other dimensions of disasters as measures and representations are deemed essential to overcome natural hazards and let people live the free life they desire or, in other words, to meet the goals of the project of modernity. This obsession with measurement is therefore firmly grounded in the Western heritage of the Enlightenment. As Horkheimer and Adorno (1947, p. 12) summarised: '*was dem Maß von Berechenbarkeit und Nützlichkeit sich nicht fügen will, gilt der Aufklärung für verdächtig*'.[1]

Of course, we recognise that numbers and quantification as well as words and writing have mattered and continue to matter in non-Western cultures (Crump, 1990; Lévi-Strauss, 1954). Rather, our concern is that the quest for pantometry and one single, ultimate truth has sustained and continues to support the hegemony of Western understanding of disaster and its extension to disaster risk reduction. It legitimises scientific paradigms and broader discourses on disasters. Lyotard (1979, p. 48) writes that '*c'est toute l'histoire de l'impérialisme culturel depuis les débuts de l'Occident. Il est important d'en reconnaitre la teneur, qui le distingue de tous les autres: il est commandé par l'exigence de légitimation*'.[2] But as Gramsci (1929–35, Q11, XVIII, §15) adds, '*credere di poter far progredire una ricerca scientifica applicandole un metodo tipo, scelto perché ha dato buoni risultati in altra ricerca alla quale era connaturato, è uno strano abbaglio che ha poco a che vedere con la scienza*'.[3] At this point, one is therefore entitled to question whether resilience can, and should be, understood.

Can resilience be understood?

In a landmark 1998 article, Mihir Bhatt asked whether vulnerability could be understood. Two decades later, the same question applies to resilience: can resilience be understood and measured? In fact, one may actually ask whether resilience *should* be understood in the first place.

Asking whether resilience can be understood raises the broader question of power and power relations in disaster studies and disaster risk reduction. In other words, whose resilience is measured, by whom, and for whom? Do those who are facing natural hazards and disasters need their resilience to be measured, especially by outsiders, and who is benefiting from such measurements? As discussed in Chapter 2, resilience remains a Western concept with a Latin origin. Hence, it hardly translates in many other languages, and attempting to capture or measure whatever it means through Western eyes may just, most often inadvertently, satisfy the appetite of the latter more than answer a local need. In a 2019 article, Bankoff (p. 229) questioned:

> it [resilience] recasts the word according to culturally-specific dictates. Depending on the context in which it is evoked, resilience either tries to restructure non-Western societies according to prescribed economic formulae or it looks for salvation in the social structures of traditional communities that it defines to its own intent.

In addition, none of the existing approaches to measuring resilience, except in the field of psychology which has been purposefully excluded from this chapter, focuses on the individual level. Hence, including those measurements of resilience at the household level entails and accepts some form of generalisation. Therefore, those people and places whose resilience is measured are seen as 'indistinguishable from one another, as controllable, homogenous objects of study who can be reduced to generalized data and explained' (Bhatt, 1998, pp. 71–72). Comparing and confronting what is otherwise heterogeneous are made possible and justified by the alleged irrefutable nature of numbers and standards in Western science. As Horkheimer and Adorno (1947, p. 29) add, '*in der Unparteilichkeit der wissenschaftlichen Sprache hat das Ohnmächtige vollends die Kraft verloren, sich Ausdruck zu verschaffen, und bloß das Bestehende findet ihr neutrales Zeichen*'[4].

In this sense, measuring resilience reflects an attempt at rationalising society through producing new, standardised knowledge, which, for Foucault (2004a), is inevitably linked to exercising power and better controlling individuals. In the context of disaster risk reduction, this means justifying the implementation of projects designed by Western institutions within a neoliberal agenda upon those at risk (Bankoff, 2019; Cheek and Chmutina, 2021; Rigg and Oven, 2015). Hewitt (1983, p. 18) once reflected that

> the vast majorities of these societies seem not to have the faintest idea of their thoroughly Malthusian condition and the natural selection that arbitrates it! Necessarily therefore, the technocrat may presume to speak for these people, but can find little value in dialogue with them or learning from them.

This contributes to reinforcing the long-lasting artificial divide between a knowledgeable, resilient world, on the one hand, and an inferior and dangerous world, on the other (Bankoff, 2001; Hewitt, 1995a). For most of its proponents, the contemporary frantic quest for enhancing measurements of resilience therefore may constitute neither more nor less than the perpetuation of the West's imperialist agenda disguised under a more appealing guise.

Notes

1 *'For enlightenment, anything which does not conform to the standard of calculability and utility must be viewed with suspicion'*. From the English edition translated by Edmund Jephcott and published by Stanford University Press in 2002.
2 *'It is the entire history of cultural imperialism from the dawn of Western civilization. It is important to recognize its special tenor, which sets it apart from all other forms of imperialism: it is governed by the demand for legitimation'*. From the English edition translated by Geoff Bennington and Brian Massumi and published by University of Minnesota Press in 1984.
3 *'To think that one can advance the progress of a work of scientific research by applying to it a standard method, chosen because it has given good results in another field of research to which it was naturally suited, is a strange delusion which has little to do with science'*. From the English edition translated by Quintin Hoare and Geoffrey Nowell Smith and published by International Publishers in 1971.
4 *'The impartiality of scientific language deprived what was powerless of the strength to make itself heard and merely provided the existing order with a neutral sign for itself'*. From the English edition translated by Edmund Jephcott and published by Stanford University Press in 2002.

5   The governmentality of disaster

On 25 January 1978, Michel Foucault (2004a) dedicated his lecture at the *Collège de France* to risks and disasters. A week before, he had lectured on food security, which was already echoing his early interest in disasters signalled in *Les mots et les choses* published more than a decade before. This little-known chapter of disaster studies, at least within the realm of this field of scholarship, was not a series of lectures on disasters for the sake of lecturing on disasters or to inform our understanding of disasters per se. Rather, Foucault was using risks and disasters as a springboard to develop his argument on governmentality that he would nail a week later during his famous lecture of 1 February.

Governmentality, Foucault writes (2004a, p. 111), is

> l'ensemble constitué par les institutions, les procédures, analyses et réflexions, les calculs et les tactiques qui permettent d'exercer cette forme bien spécifique, quoique très complexe, de pouvoir, qui a pour cible principale la population, pour forme majeure de savoir l'économie politique, pour instrument technique essentiel les dispositifs de sécurité[1].

Governmentality is a modern form of exercising power that developed in Europe throughout the eighteenth century. It was fuelled by ideas from the Enlightenment, in that it reflected concerns for an increasingly urban population, and responded to the emerging search for freedom and associated new liberal ideals. The increasing recognition that nature was posing a threat to society through extreme natural phenomena and epidemics was a key engine that justified the development of a liberal form of government that gives people an apparent sense of freedom while implicitly imposing normative rules and regulations. As Foucault (1975, p. 258) once said: '*les "Lumières" qui ont découvert les libertés ont aussi inventé les disciplines*'[2]. Mechanisms of control and discipline have since become a prerequisite to ensure people's security/safety/freedom in facing threats such as those from natural hazards.

The concept of governmentality therefore emerged in parallel to those of liberalism and disaster. As for liberalism and disaster, this new approach to government is firmly grounded in European history and the legacy of

DOI: 10.4324/9781315752167-5

the Enlightenment. It draws on three core principles of affirming sovereignty, disciplinary control and subjection, and governmental management of the population to be controlled. Imposing such a form of government is informed by the rationality of scientific evidence and requires a fine knowledge of the population through surveys and statistical data. Governmentality further builds upon specific *dispositifs*, in the sense defined in the introduction of this book, to affirm sovereignty and channel normative rules and regulations through scientific and technocratic narratives, policies, regulations, institutions and specific forms of architecture.

The influence of this modern form of government on contemporary disaster risk reduction has been immense and durable (Hewitt, 1995b). Western countries have directly built upon this heritage to design *dispositifs*[3] for disaster risk reduction that encourage normative policies and actions geared towards guiding people's behaviour and response to natural hazards in a perspective informed by Eurocentric/Western science. These *dispositifs*, including policies and actions to reduce disaster risk, have been rolled out internationally, initially through colonial forms of government and eventually through international treaties and agreements that promote the transfer of experience from Western countries to the rest of the world. As such, the Western and modern governmentality of disaster has become the international standard.

This chapter unpacks and articulates the main tenets of the governmentality of disaster at both national and international levels. It draws heavily on Foucault's approach to governmentality, especially its delineation of the concept in his lectures at the *Collège de France* between 1978 and 1979 (Foucault, 2004a,b). One may argue, as Bhabha (1994) and Young (1995a) have, that Foucault's approach to governmentality is Eurocentric and hence antithetical to our argument, which is geared to advancing a postcolonial agenda in disaster studies. It is true that Foucault's argument is Eurocentric (Lazreg, 2017). However, it is, as we discussed in the introduction of this book, because contemporary disaster risk reduction policies and actions are informed by Western discourses and principles that Foucault's approach to governmentality is particularly relevant. Indeed, our intention is to deconstruct our approach to disaster risk reduction from within.

Governmentality as the modern art of government

As discussed in Chapters 1 and 2, controlling the forces of nature to protect people and allow them to fulfil their desire for freedom has been at the core of the project of modernity (Horkheimer and Adorno, 1947). The dialectic between danger and freedom is therefore central to understanding the governmentality of disaster as we know it today. It is very much associated with the rise of liberalism as the ideology inherited from the Enlightenment. Liberalism, according to Foucault (2004b, p. 67), '*s'engage dans un mécanisme où il aura à chaque instant à arbitrer la liberté et la sécurité des individus autour de cette notion de danger*'[4].

The main objective of governmentality as a modern art of government is therefore for states to foster freedom and meanwhile ensure the safety of its people as a whole. As such, it differs from previous forms of government geared towards protecting national territories and directing the specific behaviours and lives of distinct individuals through pastoral care. It is no longer the fate of multiple distinct individuals located within the borders of the state that matters. Nor is the total number of individuals that may join the army to protect the national territory relevant anymore. It is henceforth the whole population, as a collective subject, that needs to be protected from the dangers of everyday life, including natural hazards. Governmentality is therefore a form of government of the population where power is exercised on a collective rather than on distinct individuals. This power is exercised through disciplines and control rather than force and coercion. For Foucault (1975, p. 161), disciplines are '*ces méthodes qui permettent le contrôle minutieux des opérations du corps, qui assurent l'assujettissement constant de ses forces et leur imposent un rapport de docilité-utilité*'[5].

Governmentality, as the modern art of government, builds upon the new way of knowing and managing resources which appeared during the eighteenth century and which Foucault (2004a) called political economy. Indeed, it is then that the economy, in its classic sense of governing the resources of a household for the common good, would be scaled up to the management of the state. It is the introduction of the economic in the political for the benefit of the whole population, within the broader spectrum of liberalism, that supports the modern art of government that is governmentality. The population, very much as a family, is not seen here as an aggregate of individuals, but in its overall relationship with the size of the territory, the resources available, commercial activities, diplomatic relationships, the threats to all of these, and so on. Political economy consists in scaling up an economic approach to maximising this relationship for the benefit of the entire population.

Political economy as a way of knowing and managing resources requires a fine understanding of the resources available across the territory, accurate records of commercial activities, an assessment of the surrounding threats, and detailed knowledge of the population to be sustained. It therefore draws on permanent surveillance of the resources and population to rationalise their management. The rise of political economy which underpinned governmentality as the modern art of government therefore was supported by statistics – the political science that emerged in the eighteenth century to collect and analyse data about the nature of the state. Crucial data are mostly quantitative and include, on the one hand, demographic indicators broken down by administrative areas to the household level (e.g. mortality, fertility and mean age) and, on the other hand, records of the extent of natural resources, incomes, taxes and so on, similarly disaggregated by administrative unit, or down to the household level, when relevant. This tangible evidence has provided the basis for ensuring an optimum relationship between the population and available resources in the context of diplomatic and commercial strategies and threats, such as those of epidemics and natural hazards upon which, meanwhile, science has provided increased knowledge.

To advance its political economy approach to managing a population, governmentality, as the modern art of government, required a regulatory and disciplinary *dispositif*. Foucault's concept of *dispositif* expands Althusser's earlier concept of state apparatuses. Althusser (1975) distinguished two forms of apparatuses (*appareils*) – repressive (administrations, courts and prisons, the army, the police, etc.) and ideological (schools, churches, families, the media, cultural institutions, etc.) – that allow the state to exert control over the population through verbal injunction, which Althusser called *interpellation*. Foucault's concept of *dispositif* includes both forms of Althusser's apparatuses and delineates its space of action within broader discourses. Foucault further dismisses Althusser's argument that apparatuses/*dispositifs* are organically ideological, a point to which we will return towards the end of this chapter.

The *dispositif* of security, as Foucault (2004a) calls it, is the technical instrument that facilitates the surveillance of the population and resources and their optimum management under governmentality as the modern art of government, one that manufactures a '*society of normalisation*' (Foucault, 1997). It entails giving individuals an apparent sense of freedom while implicitly imposing strong forms of control and regulation. Foucault's disciplinary *dispositif* of security is here similar to Althusser's ideological state apparatus in that they both aim at subjectifying people to the will of the state. Disciplinary *dispositifs* and ideological state apparatuses manage to guide the conduct and control of the population through multiple, invisible and non-coercive channels within the formal institutions of the state, such as schools and policies, and outside them through the media, the church and the cultural industry. In the end, both *dispositifs* and ideological state apparatuses lead the subjects to believe that their position within society and their conduct and beliefs are the natural and right ones. Here, a comparison can be drawn with Gramsci's (1971) theory of cultural hegemony and consent to the ideology of the state.

Moreover, the *dispositif* of security is a process of governmental rationalisation of the state and its resources and is geared to maintaining the right *rapport de force* between all components of the state, the *raison d'état*. As such, it is very much an exercise of policing in its broadest sense: that is, the control and management of people's activities in order to balance the forces that compose the state. The *dispositif* of security that sustains governmentality is therefore that of a state of police. The police, according to Foucault (2004a, p. 334):

> doit s'assurer que les hommes vivent et vivent en grand nombre, la police doit s'assurer qu'ils ont de quoi vivre et que par conséquent ils ont de quoi ne pas trop mourir, ou en trop grand nombre. Mais elle doit aussi s'assurer en même temps que tout ce qui, dans leur activité,

peut aller au-delà de cette pure et simple subsistance, que tout cela va bien, en effet, être produit, distribué, réparti, mis en circulation d'une manière telle que l'Etat puisse en tirer effectivement sa force.[6]

The tasks of the police are therefore to look after people and ensure that their needs are met and that they stay in good health so they can sustain the economy, commerce and trade, which the police will also help regulate. Later, we will argue that these tasks extend to reducing the risk of disaster as a threat to both people's lives and state resources.

This increasing attention given to people as part of a population and to life as a political issue led Foucault (1997, p. 216) to coin the term 'biopolitics' to describe this new art of government, which focuses on

un ensemble de processus comme la proportion des naissances et des décès, le taux de reproduction, la fécondité d'une population, etc. Ce sont ces processus-là de natalité, de mortalité, de longévité qui, justement dans la seconde moitié du 18ème siècle, en liaison avec tout un tas de problèmes économiques et politiques (...), ont constitué, je crois, les premiers objets de savoir et les premières cibles de contrôle de cette biopolitique[7].

Public health was obviously the early area of concern of biopolitics but it would soon extend to cognate mechanisms of protecting life, such as housing, insurance and savings, that will all be relevant to disaster risk reduction. In fact, governmentality would soon consider life and health in interaction with the physical environment, including the climate and hydrology, so that the biological, the economic, the political and the environmental become intrinsically linked (Foucault, 1976, 1997). The *dispositifs* of security developed to protect lives, livelihoods, and the resources of the state at large therefore will become increasingly complex and multidimensional.

In his lecture on 18 January 1978, Foucault (2004a) identified three areas of focus for these *dispositifs* of security. These areas are asserting the sovereignty of the state, regulating resources and disciplining people (the process of subjectification per se), and providing the infrastructure to organise such management of the population and resources of the state. These come together, Foucault suggests, as a triangle at the centre of which stands the population that is the main object of government: those whose lives are to be protected to ensure the prosperity of the state. It is our intention in the subsequent sections of this chapter to detail how these three areas materialise and articulate in the context of disaster risk reduction at both the national and international levels.

Asserting sovereignty through disaster risk reduction

Disasters and disaster risk reduction offer a strong opportunity for states to assert power over their territory and its constituent population and thus affirm their sovereignty. This supports what Foucault (2004a) called the theatrical practice of politics or the staging in broad daylight of the *raison d'état*. Indeed, what better opportunity to show and justify the power of the state than when it is confronted with the brutal 'forces of nature' that threaten the life of its population, as natural hazards are commonly pictured in the dominant discourse on disasters?

Controlling the forces of nature and thus protecting people's lives are therefore essential components of governmentality. Positioning natural hazards as extreme and rare phenomena outside of everyday life and the regular social fabric contributes to creating an extraneous danger that it is essential to address in name of the *raison d'état* and the collective good. In fact, natural hazards often are portrayed as enemies of the people, and of the state at large, that justify the deployment of *extra*-ordinary strategies and measures to battle against them. Reducing risk and responding to disasters then become a war, a war against nature's extremes. In the dominant discourse on disasters, we fight enemies and battle against onslaughts from the environment. We wage a war.

Waging war against natural hazards and overcoming *extra*-ordinary threats are acts of strength that reinforce the ability of the state to govern its population. Disasters allow states and governments to flex their muscles and assert power. The yearly ceremony organised by the Mexican government to remember the 1985 earthquake that struck the city of Mexico (Figure 5.1) provides an example of a state's grand display of power. In this case, it is in the context of the government's continuing struggle

Figure 5.1 Ceremony to celebrate the 28th anniversary of the 1985 Mexico earthquake, Mexico City, Mexico, September 2013

against drug rings which is expressed through the full force of the military. Similarly, the museum dedicated to the 2008 Wenchuan earthquake near Chengdu demonstrates the power of the Chinese government and its ability to overcome dramatic events as long as people follow its recommendations (Figure 5.2). The same mechanisms extend to the international level, where states send military and paramilitary search-and-rescue squads or medical/humanitarian teams (or both) to respond to disaster overseas. All of these allow foreign states to claim power and assert their strength on both the domestic and global scenes. On the other hand, those governments that cannot handle disasters with their own resources look weak in the eyes of their population. For example, a few days after the 2004 tsunami, the Indonesian government realised that it could not handle its impact on the province of Aceh with its own resources, including its otherwise-powerful military forces already located in the province. The government's subsequent call for international assistance was then seen as a sign of capitulation in the face of the 'forces of nature', one that weakened its sovereignty (Siegel, 2005).

Yet, following the 2004 tsunami, the Indonesian government had as good an opportunity as any other government to make up for their earlier submission. The process of reconstruction, and more broadly of recovery, following disasters therefore offers another wide array of opportunities for governments to strengthen their sovereignty over their territory and constituent population. The potential for creative destruction characteristic of

Figure 5.2 Exhibit in commemoration of the 2008 Wenchuan earthquake at the 5·12 Wenchuan Earthquake Memorial Museum in Mianyang, China, July 2018. Top heading reads: 'Top Communist Party officers are the primary helpers in disasters'

the project of modernity and its agenda (Harvey, 1990), discussed in the introduction of this book, here finds its ultimate expression. The prospect of rebuilding affected areas, sometimes almost from scratch, allows governments to develop new forms of urban structure and architecture, to encourage new trajectories of development, and to redistribute people in space through relocation programmes after their own desire, putting aside those of the past. In fact, reconstruction following disasters has been greatly inspired by the rebuilding of Europe following World War II. It was then essential for both victorious and defeated states to reclaim the sovereignty that had been challenged by the vicissitudes of the war (Ellwood, 1992).

Such an affirmation of sovereignty and power through disaster risk reduction, including disaster response and reconstruction, is possible only when governments have a fine knowledge of the hazards they are dealing with, the population that is exposed to them, and the extent of damage in the aftermath of disasters. The role of both data and science to sustain the governmentality of disaster is therefore essential. Science provides the evidence through palpable statistical data that allows for justified and rationalised strategies to reduce disaster risk. This data most often takes the form of deterministic or probabilistic hazard assessment, demographic indicators, and maps of both hazards and vulnerability, compiled into disaster risk assessments at different scales. In Chapter 4, we discussed how this quest for quantification also dominates the approaches to measuring the new fancy concept of resilience and its associated neoliberal discourse. This corpus of data, which is to be updated on a regular basis, facilitates surveillance and underpins the strategies of control and discipline that trickle through all types of contemporary disaster risk reduction initiatives. In fact, as Castel (1981) exposed, the prevention of any type of risk inherently requires some forms of surveillance.

The importance of tangible data to legitimise and facilitate the implementation of governmentality as the modern art of government is evident at all scales. At the international level, databases that document the occurrence of disasters, such as Emergency Events Database (EM-DAT) of the Centre for Research on the Epidemiology of Disasters (http://www.emdat.be/), have become instruments of critical importance which inform the policies of international organisations (e.g. The World Bank and The United Nations, 2010; United Nations Office for Disaster Risk Reduction, 2019b), support the advocacy of non-governmental organisations (NGOs) (e.g. Global Network of Civil Society Organisations for Disaster Reduction, 2013; Turnbull et al., 2013) and legitimise the research of scholars (e.g. Serje, 2012; Wirtz et al., 2014). Yet these databases, especially EM-DAT, are widely criticised for being skewed towards large events that trigger international humanitarian attention and that are known to a large enough number of people to stimulate research interest (Lavell, 2000; Wisner and Gaillard, 2009). In addition, available databases provide a rather rough overview of the impact of disasters as they tend to standardize data at regional and national scales. National governments are encouraged to compile their own national databases to meet the requirements of the Sendai Framework for

Disaster Risk Reduction (see below). These national databases tend to be similarly skewed towards large events and aggregated datasets, which lead to an overall generalization of the impact of disasters to the detriment of a wide range of different local realities and experiences.

At the local level, databases are replaced by disaster risk assessments that are meant to provide a picture, or table in Foucault's terms, of the local riskscape. These disaster risk assessments, informed by methodologies that we discussed in Chapter 4, are the favoured instrument for generating territorial and demographic knowledge and facilitating surveillance. Following disasters, they are replaced by post-disaster needs assessments that serve the same purpose. An example of this is the Hazard, Vulnerability and Capacity Analysis toolboxes that have flourished as part of 'community-based/led/managed' disaster risk reduction initiatives led by NGOs around the world. In some instances, these have been modified to gather for specific contexts but these toolboxes rely largely on standardised matrixes and taxonomic categorisations of hazards and factors of vulnerability and capacities (e.g. International Federation of Red Cross and Red Crescent Societies, 2007; Dazé et al., 2009). These categories usually are mirrored in quantitative and demographic indicators, each of which is one in a list of boxes to tick to complete the overall risk assessment. This approach increasingly is being replicated at the household level where families are encouraged to assess their own risk, based on hazards, vulnerability and capacities, in order to develop their personal disaster risk reduction plan (e.g. https://getthru.govt.nz/household-emergency-plan, https://www.floodtoolkit.com/).

This fine knowledge of the hazards and the population exposed to them is relevant only when it conforms to the standards of the overall discourses on disaster and their underpinning scientific paradigms. As Foucault (1997, p. 22) said: '*il n'y a pas d'exercice du pouvoir sans une certaine économie des discours de vérité fonctionnant dans, à partir de et à travers ce pouvoir*'[8]. The hazard and vulnerability paradigms respectively legitimate the war-inspired approach to disaster risk reduction and its 'community-based/led/managed' counterpart. Both the scientific narratives on disasters and the policies and actions taken to reduce risk are, as a result, mutually strengthened. Indeed, if science legitimates and thus contributes to affirming the sovereignty of the state, the state also legitimates and strengthens scientific narratives by showing their relevance to addressing real-life issues.

In fact, it is well known that the impetus for disaster studies after World War II, especially in the United States of America, was fuelled by the war industry and the American government's desire to understand how civilians would behave in case of a war. Fritz (1961, p. 653) identified two specific purposes for early disaster studies in the social sciences:

> first, to secure more adequate protection of the nation from the destructive and disruptive consequences of potential atomic, biological, and chemical attack, and second, to produce the maximal amount of disruption to the enemy in the event of a war.

He later added: 'disasters provide the social scientist with advantages that cannot be matched in the study of human behavior under more normal or stable conditions' (Fritz, 1961, p. 654). Earlier, physical scientists and engineers had already conducted studies to use natural hazards as weapons, as in the case of Project Seal which provided the governments of the United States of America, the United Kingdom and *New Zealand* with the ability to trigger artificial tsunamis to flood Japanese positions in the Pacific during World War II (Waru, 2012).

War, disaster risk reduction and the need of states for affirming and strengthening their sovereignty over both their territory and its constituent population are therefore organically linked. Gilbert (1995, p. 233), in a brilliant essay, summarised:

> disasters bear a great resemblance to war, with the causes of disasters being sought outwardly. With the concept of "agent" being used to refer to both to arms and enemies, disaster has since the beginning been explained on external grounds. As a result, human communities have been seen as organized bodies that have to react organically against aggression. (...) This paradigm holds still true, mainly because it is simple and clear. (...) The paradigm of war patterns strongly reflects the circumstances and the place where it first emerged. (...) the success of this paradigm can be explained by the nature of the demand that helped disaster studies emerge.

Discipline and regulation in disaster risk reduction

Foucault (1975) outlines four main objectives for disciplinary and regulatory mechanisms of subjection that underpin governmentality as the modern art of government. Governmentality aims (1) at controlling the body and people's behaviour in a way that Foucault called 'social orthopaedics'. It is also (2) focused on rationalising people's roles and actions through (3) a hierarchisation of tasks patterned after the organisation of the military. The regulatory nature of governmentality is finally (4) geared to control/organise space and time. Four main techniques support these four objectives: drawing tables, prescribing movements, imposing exercises and arranging tactics. In contemporary disaster risk reduction, this normative approach in its objectives and techniques is evident at all scales, from the individual and the household to the national and international scenes. It trickles through both the dominant hazard-focused and alternative vulnerability discourses.

Controlling people's bodies and behaviours has been a core focus of the disaster risk reduction agenda. In dealing with natural hazards, people are expected to behave normatively and take standardised measures guided by the recommendations of science, independently of the local context and its cultural values and economic imperatives. Disaster risk reduction seeks 'compliance behavior' (Perry, 1994). For all hazards, there are, therefore, appropriate and inappropriate behaviours and measures such as 'move uphill' and 'don't drive' when dealing with floods, 'run uphill' and 'don't head towards the sea' in the event of a tsunami, and the most celebrated 'drop, cover and hold' during an earthquake. Agreeing on one single standard recommended behaviour (i.e. the prescribed movements in Foucault's terms) thus is critical and stirs significant debates amongst scientists, as illustrated in the discussions that divided 'experts' on what to do when the earth shakes (Mahdavifar et al., 2009; Spence et al., 2011).

These normative behaviours and actions are central to the so-called education campaigns that underpin disaster risk reduction programmes all over the world. These feature in both technocratic hazard-driven and community-based/led/managed initiatives. Education is here a one-way street where 'experts' tell local people what to do in order to normalise their behaviour in the same way as orthopaedics. They draw on the assumption that people do not know what to do and do not spontaneously take the appropriate measures when dealing with hazards and disasters. Rancière (1987, p. 44) elaborates on this process of educating those who allegedly do not know:

> il y a des inférieurs et des supérieurs; les inférieurs ne peuvent pas ce que peuvent les supérieurs. La vieille [approach to education] ne connait que cela. Il lui faut de l'inégal, mais non point de cet inégal qui avoue le décret du prince, de l'inégal qui va de soi, qui est dans toutes les têtes et dans toutes les phrases[9].

Education campaigns for disaster risk reduction (i.e. one form of Foucault's tactics) thus reflect Rancière's old approach to education ('*la vieille*') which Freire (1968, II, p. 2) called a banking approach, where '*o "saber" é uma doação dos que se julgam sábios aos que julgam nada saber. Doação que se funda numa das manifestações instrumentais da ideologia da opressão – a absolutização da ignorância*'[10]. In disaster risk reduction, this alleged divide in knowledge and ability between the experts – whether they are scientists, government officials or NGO staff – and the local people justifies a blunt transfer of knowledge.

It is the positioning of the norm as absolute truth dictated by Western science and its constant reiteration through diverse channels of communication – including education campaigns – that will make discourses on disaster so widely accepted and durable. The normative nature of education campaigns is thus the broader extension through society of the hegemony of Western academic knowledge in understanding disasters. They contribute to making recommended behaviours common sense in Gramsci's (1971) terms or evidence in Althusser's (1975) to subjectify people through intellectual consent (Rancière, 1987) and corporal compliance (Foucault, 1975). As such, expected behaviours and actions at the time of a disaster are not seen as constraints imposed upon them by powerful and coercive institutions. They are embedded within broader and hegemonic discourses on disaster. As Althusser (1975, p. 121) once said: in the end, '*les sujets "marchent tout seuls"*'[11].

As a result, recommended and normative behaviours are featured in posters, brochures, media adds, cartoons, magnets, comics and so on. Sometimes, they are wrapped with a tokenistic cultural packaging to make it somehow relevant to the local context. Signs are also set up in hazard-prone areas to provide a constant reminder of what to do and to guide people's behaviours. Such is the case with tsunami warning and evacuation signs, for which scientists around the world have agreed upon a standardised iconography and colouring, whether they are to be set up in Samoa, the Philippines, Martinique, the United States of America or Chile (Intergovernmental Oceanographic Commission, 2020). The ultimate form of control over people's bodies and actions when dealing with natural hazards is, however, the conduct of drills where individuals are expected to learn and enact how to behave. Drills (which Foucault would call exercises) are conducted as part of both hazard-driven and technocratic initiatives and 'community-based/led/managed' approaches to reduce disaster risk (e.g. https://www.shakeout.org/, International Federation of Red Cross and Red Crescent Societies, 2012).

It is through drills that '*les rapports de pouvoir passent à l'intérieur des corps*'[12] (Foucault, 1977b, p. 4), where norms get literally embodied by people who face natural hazards. During disaster drills, participants are trained to recognise and respond to warning signals in a way that is crucial to sustain the disciplinary nature of disaster risk reduction. As Foucault (1975, p. 195) wrote:

> toute l'activité de l'individu discipliné doit être scandée et soutenue par des injonctions dont l'efficacité repose sur la brièveté et la clarté; l'ordre n'a pas à être expliqué, ni même formulé; il faut et il suffit qu'il déclenche le comportement voulu[13].

During drills, participants also 'learn' how to play specific roles as part of a larger distribution of responsibilities within a particular collective, whether in an administrative area or a so-called 'community'. This process is believed to contribute to rationalising collective response and maximising local skills and resources while making sure that those most in need are looked after by kin and neighbours in the 'community-based/led/managed' approach. This often leads to a hierarchisation of tasks, in which some people or stakeholders take a leading position and organise the overall response on behalf of others, as in a military chain of command. This process of regulating and rationalising how to deal with natural hazards appears in both the technocratic emergency management system, patterned after military structures, and 'community-based/led/managed' disaster risk reduction initiatives where local people are distributed tasks to perform in order to contribute to the overall good of their 'community'.

This regulation and rationalisation of tasks and roles usually are determined for specific areas and require a *quadrillage* of the territory, which reproduces on a broader scale the principles and structure of the panopticon that

Foucault (1975) placed at the core of his argument in *Surveiller et Punir*. This *quadrillage* is hierarchical and plays out on different scales from the national to the very local level, hence matching the tasks and responsibilities of different levels of institutions within the government and other stakeholders. It therefore allows for a systematic and thorough control and surveillance of people and their response to natural hazards and disasters to be embedded within imbricated pyramids of responsibilities dominated by experts and the most powerful stakeholders at every rung of the structure, from village/neighbourhood leaders to heads of state and experts who advise them.

Optimum *quadrillage* of the territory to maximise the *rapport de force* requires a fine knowledge of the spatial dimension of hazards, vulnerability and capacities/resilience. Therefore, maps (one essential dimension of Foucault's tables) are considered crucial for disaster risk assessment and reduction, whether framed from the technocratic hazard perspective or the 'community-based/led/managed' approach (e.g. Alexander, 2002a; Mercer et al., 2009). Hazard maps allow the delineation of areas where actions are to be taken. Mapping vulnerability, in all its multiple dimensions, is also deemed necessary to assess what and who need the most attention. In parallel, mapping out capacities allows the availability of local resources, skills and knowledge to be assessed and local response to be maximised. Such maps build upon and complement the surveys and other demographic data discussed in the previous section. They are drawn on multiple scales in the same process of systematic and thorough assessment (i.e. the broader dimension of Foucault's tables), surveillance and control of the territory and people described above. This process is considered essential for the effective planning of actions to reduce disaster risk. Usually, hazard, vulnerability and capacities maps are ultimately integrated in risk maps that allow for a supposedly accurate assessment of the relationship between needs and resources across scales.

Maximising the discipline and control of people's behaviour, the hierarchisation of tasks, and the *quadrillage* of the territory finally requires an approach that fragments time into a series of actions. As a result, disaster risk reduction has long been seen as a sequence of initiatives captured in the so-called disaster management cycle or continuum, where prevention and mitigation precede the preparedness that comes ahead of response, while recovery is supposed to come last, after an 'event' has occurred (Alexander, 2002a; Coetzee and van Niekerk, 2012; Quarantelli, 1990). Each and every stage of the cycle, or continuum, is further broken down into sub-sequences, such as in the recovery stage, which usually includes rehabilitation, followed by reconstruction and development (Haas et al., 1977), with the same goal of fragmenting time in order to better control and rationalise actions in the context of available needs and resources. Despite widespread criticisms (e.g. Neal, 1997; Kelly, 1998–99; O'Brien et al., 2006), the disaster management cycle, or continuum, has been influential and durable for the very reason that it epitomises governmentality in its ambition to rationalise government in time and space to better normalise and control actions and behaviours.

The *dispositif* of disaster risk reduction

A *dispositif* is a complex and multidimensional apparatus that includes scientific narratives, discourses, institutions, policies and regulations, funding mechanisms, and specific forms of infrastructure and architecture, designed in order to strategically respond to a specific issue (Foucault et al., 1977). There is such a *dispositif* for disaster risk reduction. We have already exposed that it is sustained by the scientific paradigms and broader discourses that (1) reflect the hegemony of Western ontologies and epistemologies in disaster studies and (2) allow for normative disciplinary and regulatory actions to reduce disaster risk. In this section, we turn to the mechanisms through which paradigms and discourses result in normative actions and strategies. We also stretch Foucault's concepts of *dispositif* and governmentality designed at the level of the state to international mechanisms of government. We further seek inspiration from Althusser (1975) to differentiate governmental mechanisms specifically dedicated to disaster risk reduction from other channels through which the regulatory nature of governmentality is imposed upon people, especially those at risk of disaster.

Global arrangements and the hegemony of international agreements

The normative, disciplinary and regulatory strategies are celebrated, standardised and imposed upon all stakeholders of disaster risk reduction through international agreements and guidelines (Revet, 2018). A fine-grained analysis of the international agreements for disaster risk reduction signed by countries since the mid-1990s (i.e. the Yokohama Strategy and Plan of Action for a Safer World of 1994, the Hyogo Framework for Action of 2005, and the Sendai Framework for Disaster Risk Reduction of 2015) would show that, even if the contents have changed and marked a slight shift of attention from hazards to vulnerability, they all still promulgate normative discourses based on Western understandings of disasters. Another example is the Sphere Standards that provide guidance for humanitarian workers in the aftermath of disasters. These explicitly normative standards are informed by the Western principles of human rights that were inherited from the Enlightenment and that are meant to apply all over the world:

> The standards are informed by available evidence and humanitarian experience. They present best practice based on broad consensus. Because they reflect inalienable human rights, they apply universally. However, the context in which a response is taking place must be understood, monitored and analysed in order to apply the standards effectively.
>
> (Sphere Association, 2018, p. 6)

These international agreements and guidelines, despite their non-binding nature, leave little space to accommodate alternative, non-Western approaches because they entail reporting mechanisms and strings with regard to funding opportunities.

For example, the Sendai Framework for Disaster Risk Reduction strongly suggests that governments follow strict guidelines with regard to the design of their national policies so that they meet a number of targets reflected in specific indicators. This monitoring framework, as it is called, is an obvious instrument of surveillance and control directly framed after Western understandings of disasters and based on criteria set against the experience of Western countries in reducing what they call disasters. In fact, there is now an official Scientific and Technical Advisory Group that advises the United Nations Office for Disaster Risk Reduction on implementing the Sendai Framework for Disaster Risk Reduction. The Sendai Framework for Disaster Risk Reduction, like its predecessors, further guides the priorities of international organisations and donors that provide financial supports to government, NGOs and the other stakeholders of disaster risk reduction (Revet, 2018). As a result, the policies designed by national governments, and the actions taken by a diverse range of stakeholders, are largely steered by, and accountable to, these international priorities and the discourses that underpin them. In the next two chapters of this book, we focus more specifically on the discourse on climate change adaptation and the discourse that encourages inclusion in disaster risk reduction at large.

Attached to these international agreements, especially to the Sendai Framework for Disaster Risk Reduction, are funding mechanisms that serve as powerful instruments of control and regulation on all scales but with particular normative impacts at the local level (Revet, 2018). International funding schemes, whether from the United Nations agencies, development banks or wealthy government donor offices, are indeed indexed after the priorities of the Sendai Framework for Disaster Risk Reduction as well as those of the Sphere Standards and other more specific agreements. Funding schemes therefore define priorities that reflect Western discourses on disasters, whether they reflect either side of the nature/ hazard–culture/vulnerability and encourage normative initiatives that not only reflect disciplinary and regulatory mechanisms to address disaster risk but also support the imperialist agenda of the West. That is, they encourage the transfer of experience, technology and funding from affluent to less wealthy countries. Beneficiaries of these funding schemes, which are mainly government agencies and NGOs and other civil society organisations, have no choice but to bow down to the terms of the funding schemes should they wish to implement initiatives that, in the end, allow them to meet the requirements of the Sendai Framework for Disaster Risk Reduction and other international agreements (Gibson, 2019).

It is also not unusual for funding schemes dedicated to government agencies geared to reducing disaster risk, or to recovery following disasters, to come with strings that reflect, or are part of, broader structural adjustment programmes imposed by the International Monetary Fund, the World Bank and the states that support these two organisations. These notably include the removal of quotas and tariffs on imports, deregulation of the economy, and the privatisation of basic services. These measures allow for the spreading and enforcement of neoliberal principles on affected and frequently strangled states. The most famous case is probably Haiti following the 2010 earthquake (Schuller and Moralles, 2012) but similar strategies occurred in the aftermath of Hurricane Katrina in the United States of America in 2005, Hurricane Mitch in Honduras in 1998, the 2001 earthquake in Salvador, the 2004 tsunami in Indonesia, and most recently following the 2017 hurricane season in the Caribbean, especially in Puerto Rico (Barrios, 2017; Bello, 2006; Bonilla, 2020; Gunewardena and Schuller, 2008; Wisner, 2001).

The governmental infrastructure and mechanisms

In line with the recommendations of international agreements such as the Sendai Framework for Disaster Risk Reduction, most countries these days have some sort of national, and frequently subnational, disaster risk reduction policies, guidelines and strategies (United Nations Office for Disaster Risk Reduction, 2019b). Their actual labelling, ranging from emergency management and civil defence to disaster risk reduction policies, reflects how far across the binary national attention has shifted from nature/hazard to culture/vulnerability. In practice, though, how far the national discourse on disasters has shifted has not really altered the overall assumption that the entire country and all administrative subdivisions and sometimes households need a plan to reduce disaster risk and these plans need to be guided by some sort of expertise, whether it is focused on better understanding hazards or on fostering the inclusion of those most vulnerable.

The importance of planning for disaster risk reduction, as briefly discussed in the introduction of this book, is a very powerful instrument of normalisation and control, which reflects the art of governmentality inherited from the Enlightenment. Quarantelli (1981, p. 1) once explicitly stated that 'all planning should be future oriented. A future orientation is necessary both in thinking about disasters generally and about specific disasters. Too often in thinking in the present we are prisoners of the past'. Yet this Western approach to planning for the future is a Western idea that does not translate into all cultures, especially where decisions are made on an *ad hoc* basis. In fact, 40 years ago, Quarantelli (1980, p. 1) himself was already concerned about the normative nature of planning, especially beyond the Western world:

> Research has demonstrated that much preparedness activity is misguided in that the planning assumes people should adjust to plans rather than the converse. There is also an unfortunate tendency to conceive of preparedness as primarily the drawing up of written plans; whereas, research indicates that the production of plans should be only a minor of the process.

However, having a disaster plan has become a central component of disaster risk reduction policies (e.g. Alexander, 2002a; Drabek, 1986; United Nations Office for Disaster Risk Reduction, 2019a). These plans are the direct result of disaster risk assessments and associated mechanisms of surveillance that allow for a multi-layered control of the population, from the national to the household scale. They are the instrument that allows the rationalisation of time and space, as discussed in the previous section. These plans often are crafted after normative templates which mirror the importance of Western concepts in understanding disaster, whether the approach is technocratic and hazard-focused or 'community-based/led/ managed' and vulnerability-focused (Figure 5.3).

With plans usually come guidelines on how to implement them, as well as codes to legalise some of the recommended measures, thus adding further regulations and normative procedures, which in the end leave very little space for alternative approaches. Building codes have been a very powerful conduit for regulating and normalising disaster risk reduction. These codes are designed after engineering standards that draw on Western science rather than traditional/indigenous construction techniques and architectures (Lewis, 1976b, 2003). The hegemony of Western science has made these building codes almost unchallengeable, so that the lack of such codes or the inability of a government to apply them is often identified as a major cause of disaster (Lewis, 2008).

International agreements such as the Sendai Framework for Disaster Risk Reduction also recommend that these plans be implemented by dedicated agencies or institutions. The overall structure of these agencies is

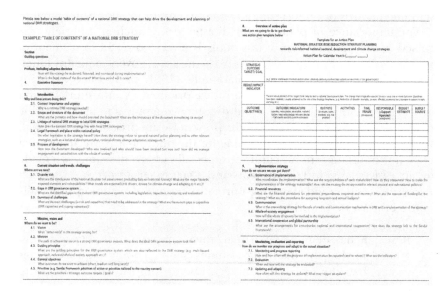

Figure 5.3 Exemplar of template for designing a national disaster risk reduction plan

relatively similar wherever in the world they occur, despite the nuances in their actual arrangement, staffing, and funding mechanisms, and are all templated after Western standards (Lavell et al., 2012). In fact, in some regions of the world, regional organisations have recommended and supported national governments in following the same framework and structure. For example, the Pacific Community has steered the strengthening of national policies and agencies across the region following a similar template that follows Western standards, rather than local understandings, of what the West calls disaster and traditional forms of government (Pacific Islands Applied Geoscience Commission, 2009).

The pattern is also for both policies and agencies tasked with reducing disaster risk to follow a hierarchical structure, from the national level down to subnational and local administrative subdivisions, that supports the hierarchical distribution of tasks amongst stakeholders of disaster risk reduction discussed in the previous section. In most countries, this hierarchical structure reflects a military heritage of waging a war against natural hazards and provides the opportunity for governments to affirm their sovereignty when dealing with disasters. The military structure, in fact, has informed governmentality as the modern art of government. Foucault (1975) described well how the military structure in the eighteenth century was considered the most elaborate and efficient institution for control and regulation. This is still reflected in disaster risk reduction where, it does not matter whether or not the national discourse on disaster has shifted its attention to vulnerability, it is still about being able to protect people and save lives. It is about biopolitics. In this perspective, dominant strategies to reduce the risk of disaster do mimic strategies of war with multiple battles every time a flood or an earthquake occurs.

Such strategies require the special forces and resources of the military and follow top-down chains of command that fit nicely within military structures or those of its civilian arms known as civil defence or civil protection (Alexander, 2002b). Military or paramilitary *dispositifs* are the norm in dealing with disasters. By extension, everyday actions to reduce disaster risk participate in a police state. Hence, not only contemporary police forces but also their counterparts in uniforms look after disasters. These include fire brigades, civil protection teams, paramilitary organisations, scout groups and other security agents. Police and the other organisations in uniforms are tasked with supervising people who deal with natural hazards. People who (as discussed in the previous section) are deemed unable to face disasters on their own because they lack appropriate knowledge or because their behaviour is expected to be inappropriate. Police and other officers in uniform are therefore those who are meant to steer evacuation procedures as well as to prevent looting and other crimes, even though it is well known that the latter seldom if ever occur in a time of disaster (Quarantelli and Dynes, 1972). To support these police actions, again in a broad sense, these organisations and agencies rely upon their own mechanisms of data collection (i.e. often called intelligence, after the military vocabulary) and surveillance.

The other dimensions of the dispositif of disaster risk reduction

The hierarchical and military-inspired nature of disaster risk reduction policies and agencies is the ultimate and explicit expression of the normative, disciplinary and regulatory approach to addressing what the West calls disaster. Yet there are other, often less explicit, pathways to convey a normative approach to reducing disaster risk. One is the school, which Foucault (1975, 1976) flagged as one of the two institutions (along with the military) where the disciplinary and regulatory nature of governmentality as the modern art of government first developed. It is at school that the most powerful forms of social orthopaedics happen from a very young age. At school, children learn not only what a disaster is but also how to respond to them through standardised drills in dealing with fire, earthquake or whatever other local hazards exist. In fact, the school is considered one of the most powerful channels for instilling a culture of risk/ disaster among children but also in the hope that it will have a ripple effect on their households. '*Teaching disaster risk reduction*' is meant to be a pathway towards behavioural change – or social orthopaedics, in Foucault's terms – in accordance with the latest guidelines from the United Nations Office for Disaster Risk Reduction (2020, p. 60): 'The research showed that child-adult transmission of risk reduction information to increase household safety has great potential to mobilise adults and catalyse behaviour change'.

This discourse and expectation are also those of many civil society organisations, especially international NGOs, advocating for the inclusion of children and other groups labelled vulnerable after Western standards. As we discussed in the previous section, the priorities of NGOs and the nature of their actions on the ground are dictated largely by the terms of available funding mechanisms and the objectives of international agreements such as the Sendai Framework for Disaster Risk Reduction (Gibson, 2019). In Chapter 7, we show that the approach of many NGOs and their partners is therefore necessarily normative in the sense that it imposes a Western discourse around inclusion and participation with usually limited opportunities for considering local cultural values. Inclusion and participation are, in fact, often directed towards a planning process similar to that of governments with rigid templates offered by international NGOs to their local partners. As such, they often perpetuate an agenda that contributes to the rolling-out of Western discourses on disasters around the world (Revet, 2007, 2018; Schuller, 2012).

There are many other, less obvious pathways that convey Western discourses on disasters and indirectly shape both people's behaviours and the responses devised to reduce disaster risk. It is not our intention here to provide an exhaustive analysis of these as they would require a book of their own. We only endeavour to show the diversity of the channels through which Western discourses and disciplinary and regulatory mechanisms pervade all dimensions of society.

One of these pathways is the media, both traditional and social. A considerable literature discusses how the media has been instrumental in spreading myths about disasters (e.g. Alexander, 2014; Scanlon, 2007; Tierney et al., 2006). Myths about the significance of *extra*-ordinary hazards and also about people's response to disasters and how these myths about irrational and antisocial behaviours have justified the intervention of government agencies, including the military, the police and NGOs. These assumptions about the *extra*-ordinary nature of disasters and how people respond to them have proved durable in popular culture (Quarantelli and Davis, 2011; Webb, 2007).

They have been reinforced by the portraits that films (especially Hollywood movies), cartoons, comics, novels, songs and other forms of art have drawn of disasters (Alexander, 2012; Quarantelli, 1980b; Quarantelli and Davis, 2011). Films, novels, songs and other artworks have further contributed to supporting the imperialistic agenda of the West by spreading and perpetuating a skewed and homogenous representation of the non-Western world as dangerous and unable to deal with disasters on its own, hence justifying the intervention of international organisations, NGOs and scientists to rescue vulnerable locals. In fact, one could easily apply the approach of JanMohamed (1985), Spivak (1987) or Said (1978) to studying what the last of these calls Orientalism to the study of disasters in the non-Western world. Emerging studies of games, an increasingly popular tool for disaster risk reduction, also show that the majority of board games, card games and puzzles used to foster disaster risk reduction emphasise the importance of *extra*-ordinary hazards that overwhelm vulnerable people who are unable to deal with disasters on their own (Culibar, 2020). Mainstream and serious video games that portray disasters convey the same message and contribute to spreading a skewed image of disasters which reflects the dominant Western discourses (Gampell et al., 2020).

Museums and memorials are less studied but nonetheless powerful conduits for conveying a particular image of disasters which supports Western discourses and justifies the intervention of institutional actors to rescue those who are affected. Disaster museums are particularly credible to their audience, as they convey what is believed to be the truth, supported by undeniable scientific facts and evidence. Indeed, it is the norm for museums that commemorate disasters, whether located in Europe, the United States of America, China, Japan, the Philippines, Indonesia, Australia or Aotearoa, to guide visitors through a retrospective journey that most often starts with a science-based exhibit of the *extra*-ordinary natural phenomenon involved (Figure 5.4). Many museums also include recommendations on how to deal with future natural hazards and disasters, emphasising the normative recommendations that we discussed in the previous section of this chapter (Figure 5.5). Finally, it is not rare for museums that commemorate earthquakes to include a shake room/house which simulates a tremor and encourages people to drop, hold and cover, thus allowing normative recommendations to penetrate the visitors' bodies (Figure 5.6).

Figure 5.4 Science-based exhibit on volcanoes at the Museum Gunungapi Merapi in
Sleman, Indonesia, July 2009

On the durability of the governmentality of disaster

The overall architecture of the *dispositif* of disaster risk reduction has been
relatively stable and durable since the eighteenth century and the emergence
of governmentality as the new form of government. Its durability obviously
stems from the hegemonic nature of both Western knowledge in understanding disaster and the overall project of modernity it contributes to. For those
in power, governmentality and its *dispositif* are also an efficient, cost-effective
and politically rewarding approach to governing a territory and its constituent population (Foucault, 2004a). It is therefore no surprise that for the past
40 years the governmentality of disaster has built upon the injunction that
'*prevention is better than cure*' (Davis, 1984; Hagman et al., 1984) and that

> reducing losses and damages in the event of a disaster is often the key
> motivating factor for disaster risk management [DRM] (…). However,
> even if the anticipated disaster does not occur for a long time, increased
> resilience means that background risk is reduced and economic

Figure 5.5 Normative recommendation on how to respond to an earthquake at the California Academy of Sciences museum in San Francisco, United States of America, August 2013

Figure 5.6 Earthquake and volcano shake house at Tāmaki Paenga Hira (Auckland War Memorial Museum), Aotearoa, November 2010

development potential unlocked (...). In addition to these primary objectives of DRM, investments in resilience may yield further social, economic and environmental co-benefits (...). In the medium- to long-run, these benefits can trigger a wide range of benefits across society, income groups, geographic regions, government entities, industries, and supply chains.

(Tanner and Rentschler, 2015, p. 10)

This does not mean that the *dispositif* of disaster risk reduction has remained in exactly the same shape over the past four decades. The changing global political and economic environment and the rise of neoliberalism in the 1980s have marked a progressive withdrawal of the state for the benefit of individualistic values. Thereafter, freedom has been not only about being protected from natural hazards but also for individuals to make free and informed decisions about whatever concerns them most, a priority that has supported the discourse on inclusion and participation in 'development' at large and disaster risk reduction in particular. Both the withdrawal of the state and the need to foster inclusion and participation have sustained the strengthening of NGOs and other civil society organisations in disaster risk reduction as well as in other dimensions of society (Bankoff, 2019).

The rise of neoliberalism, as a new form of liberalism and associated values, has facilitated the progressive shift of attention across the binary, from nature/hazard to culture/vulnerability, observed in disaster risk reduction policies and actions over the past 30 years (Bankoff, 2019). The rise and progressive skewing of the vulnerability paradigm in disaster studies, discussed in Chapter 2, and, eventually, of its resilience/capacities spin-off and all the individualistic values the latter carries, have supported and justified this transfer of attention. This shift has further required a downscaling of the overall focus of policies and actions from the population as a whole, as in Foucault's framing, to 'communities' and their resilience.

Unsurprisingly, this seemingly radical transfer of attention across the binary is no more significant in policy and practice than it is in scholarship. As we discussed in Chapter 1, the governmentality of disaster and its *dispositif* remain within the bounds of the nature/hazard–culture/vulnerability binary and its Western ontological and epistemological legacy. Three key dimensions of disaster risk reduction policies and actions have survived the shift across the binary with very little reconsideration and this shows how limited the impact of the paradigm shift has been in policy and practice. As such, they continue to sustain an approach to disaster risk reduction that mirrors the modern art of government of the West that is governmentality.

First, downscaling the focus from the population at large to local 'communities' seems tokenistic in that 'communities' are, in practice, often (falsely) perceived as homogenous entities (Cannon, 2014), very much like the wider population was, and still is, considered. In Chapter 7, we show that, in general terms, the practice of participatory disaster risk reduction has not yet lived up to the promises of its pluralistic ethos, thus mirroring the unfulfilled potential of the paradigm shift in disaster studies that we discussed in Chapter 2. As Edwards and Hulme (1995, p. 13) were already observing more than two decades ago,

> the type of appraisal, monitoring and evaluation procedures insisted on by donors, especially their heavy reliance on "logical framework" approaches and bureaucratic reporting, may also distort accountability by overemphasising short-term quantitative targets, standardising indicators, focusing attention exclusively on individual projects or organisations, and favouring hierarchical management structures.

Second, planning remains at the centre of disaster risk reduction policies and actions, whether they are hazard-driven or vulnerability/resilience/capacities-informed, technocratic or 'community-based/led/managed'. In Chapter 1, we signalled that planning is the very essence of the quest for rationalisation that the Western project of modernity carries. It epitomises Foucault's (2004b) articulation of the freedom–control dialectic as it is a direct form of guiding and controlling people's future while upholding the promise for progress and freedom.

Third, planning for disaster risk reduction, from whichever perspective, is still to be informed by evidence which preferably is both quantitative and statistical because these are believed to provide the truth and to facilitate accountability (see Chapter 4). Similarly, the implementation of plans is to be monitored and assessed according to (quantitative) criteria and indicators set as part of the planning process. Thus, by nature, planning, as a political endeavour, has to be guided by an economic approach, so that (1) the actions to be taken reflect the needs and available resources and (2) the relationship between the needs and resources be monitored. It is still about policing disaster risk reduction.

The persistence of these three key tenets of governmentality mirrors the continuing legacy and prominence of Western *métarecits* of disasters and approaches to reducing disaster risk. One may argue that this approach has achieved some success both in the West and elsewhere in the world, as shown by the declining number of disasters recorded by international databases and headlined by many organisations to legitimise their strategies and initiatives (e.g. United Nations Office for Disaster Risk Reduction, 2019a). However, we have seen earlier that these databases are skewed, not only towards large events but also towards what the West understands a disaster is. The databases are therefore an integral part of the discourses and *dispositif* that they legitimate and, as such, cannot be taken as an objective measure of success should such a measure be needed in the first place.

Lyotard (1979, p. 71) once suggested that '*l'argumentation exigible pour l'acceptation d'un énoncé scientifique est donc subordonnée à une "première" acceptation (…) des règles qui fixent les moyens de l'argumentation*'[14]. In the context of disasters and disaster risk reduction, both the rules and argument are skewed towards one truth, which is that of the West, informed by the legacy of the Enlightenment. As Lyotard (1979) concluded, the problem then becomes about proving or challenging the proof. This question is at the core of the argument of this book, which is geared at challenging the rolling-out and relevance of a Western approach to government onto regions of the world that do not have the same heritage.

Power and knowledge in disaster risk reduction

The overall argument of this book is that the normative and regulatory nature of the governmentality of disaster around the world is possible only because of the hegemony of Western discourses on disasters. Indeed, the ontological and dialectical relationship between nature/hazard and culture/vulnerability, which underpins the *métarecits* of disaster as delineated by disaster studies, emulates the dialectical relationship between freedom and control that informs governmentality as the modern art of government. Foucault (2004b, p. 69) once wrote about governmentality that

> c'est l'apparition (…), dans ce nouvel art de gouverner, de mécanismes qui ont pour fonction de produire, d'insuffler, de majorer des libertés, d'introduire un plus de liberté par un plus de contrôle et d'intervention. C'est-à-dire que là, le contrôle n'est plus simplement (…) le contrepoids nécessaire à la liberté. C'en est le principe moteur[15].

Therefore, protection and security, to use Foucault's terms, justify some form of intervention (control) from governments, international organisations and other stakeholders such as the NGOs, guided by the scientific evidence deemed essential to understand and deal with *extra*-ordinary phenomena and using appropriate resources. Disasters thus become powerful, convenient and justified conduits for implicitly controlling people and establishing mechanisms of surveillance, whether these reflect a liberal or, these days, neoliberal ideology.

We therefore argue, against Foucault, that the *dispositif* of disaster risk reduction, across all scales (from the local to the international) and across all its dimensions (institutional, legal, financial, scientific, architectural and cultural), is inherently ideological. Liberalism, at the core of Foucault's (2004b) argument on governmentality, and neoliberalism are ideologies if we follow Althusser's (1975, p. 97) definition of the latter, which is a '*système des idées, des représentations qui domine l'esprit d'un homme ou d'un groupe social*'[16].

Not all states around the world adhere to the same liberal or neoliberal ideologies, and we recognise the existing diversity of approaches to government. Foucault's (2004b) argument, upon which we draw our analysis here, is explicitly and deliberately based on countries where the art of governing has the same Western heritage, especially in Europe and Northern America. Nonetheless, this is where the international discourses and *dispositif* of disaster risk reduction are clearly grounded and from which they are disseminated to other regions of the world. They are imposed upon governments through international agreements, such as the Sendai Framework for Disaster Risk Reduction, and colonial and neocolonial heritages in contemporary forms of government that do not necessarily reflect traditional approaches. As such, national *dispositifs* of disaster risk reduction, as imposed by international agreements and supported by Western discourses on disasters, are instruments for the West to spread its hegemonic and misleadingly diverse approaches to disaster risk reduction and, by extension, a liberal/neoliberal agenda. Disaster risk reduction thus appears to be an instrument of imperialism for the West (Middleton and O'Keefe, 1998). An instrument that contributes to 'the construction of consent' to neoliberalism (Harvey, 2005).

The explicit rationale for this imperialist agenda is simple and powerful. A report, among many documents of the same kind, commissioned by United Nations agencies, the World Bank's Global Facility for Disaster Reduction and Recovery, and Oxfam (2010, p. 4), a representative range of stakeholders of disaster risk reduction at the international level, puts it like this: 'unchecked by the integration of risk into development, the impact of disasters will grow and grow. Development must be risk-proofed now, so as to prevent massive losses of life, livelihoods and growth in the future'. In other words, disasters, in Western discourses, prevent people from flourishing and living a free life. They are a humanitarian issue that cannot be left unattended, and it is a task for the Western project of modernity to rescue the world using its reason, science and technology inherited from the Enlightenment (Revet, 2018). As Bankoff (2001, pp. 26–27) adds,

> the popularisation of this representation through the mass media also generates a moral obligation on behalf of Western nations to employ their good offices to 'save' these vulnerable populations from themselves and to render the regions they inhabit safer for investment and tourism.

This is the Western heritage that shapes the governmentality of disaster at the national level, which consequently is hegemonic in nature. The normative nature of the disciplines and regulations promoted by national *dispositifs* designed to reduce the risk of disaster is set to convey the dominant liberal/neoliberal ideology across broader society. An ideology that draws on the absolute truth provided by Western science and that justifies the state and other actors coming to support vulnerable people, who apparently do not know how to deal with or have the resources to deal with natural hazards, through rational mechanisms and instruments such as risk assessment, education, drills and planning. The multiple, powerful and durable pathways through which the Western discourses on disaster are conveyed into people's minds and bodies are devised to ensure subjection to the dominant views on disaster, whether they are hazard-centric or vulnerability-driven, and their reproduction in time. By extension, initiatives to reduce disaster risk contribute to the subjection of people around the world to the norms and values of liberalism and neoliberalism. In this perspective, disaster risk reduction is no different from medicine, development, environment sustainability or democracy (Bankoff, 2003a; Escobar, 1995; Middleton and O'Keefe, 1998). We provide two examples of such processes in the subsequent chapters.

Notes

1 '*The ensemble formed by institutions, procedures, analyses and reflections, calculations, and tactics that allow the exercise of this very specific, albeit very complex, power that has the population as its target, political economy as its major form of knowledge, and apparatuses of security as its essential technical instrument*'. From the English edition translated by Graham Burchell and published by Palgrave Macmillan in 2007.

2 '*The 'Enlightenment', which discovered the liberties, also invented the disciplines*'. From the English edition translated by Alan Sheridan and published by Vintage Books in 1995.

3 We use '*dispositif*' (singular) when we refer to the overarching and international scale and '*dispositifs*' (plural) in reference to the national level.

4 '*turns into a mechanism continually having to arbitrate between the freedom and security of individuals by reference to this notion of danger*'. From the English edition translated by Graham Burchell and published by Palgrave Macmillan in 2008.

5 '*These methods, which made possible the meticulous control of the operations of the body, which assured the constant subjection of its forces and imposed upon them a relation of docility-utility*'. From the English edition translated by Alan Sheridan and published by Vintage Books in 1995.

6 '*must ensure that men live, and live in large numbers; it must ensure that they have the wherewithal to live and so do not die in excessive numbers. But at the same time it must also ensure that everything in their activity that may go beyond this pure and simple subsistence will in fact be produced, distributed, divided up, and put in circulation in such a way that the state really can draw its strength from it*'. From the English edition translated by Graham Burchell and published by Palgrave Macmillan in 2007.

7 '*a set of processes such as the ratio of births to deaths, the rate of reproduction, the fertility of a population, and so on. It is these processes—the birth rate, the mortality rate, longevity, and so on—together with a whole series of related economic and political problems (…) which, in the second half of the eighteenth century, become biopolitics' first objects of knowledge and the targets it seeks to control*'. From the English edition translated by David Macey and published by Picador in 2003.

8 '*Power cannot be exercised unless a certain economy of discourses of truth functions in, on the basis of, and thanks to, that power*'. From the English edition translated by David Macey and published by Picador in 2003.

9 '*there are inferiors and superiors; inferiors can't do what superiors can. The Old Master knows only this. It depends on inequality but not the inequality that acknowledges the Prince's decree, the inequality that goes without saying, that is in all heads and in all sentences*'. From the English edition translated by Kristin Ross and published by Stanford University Press in 1991.

10 '*knowledge is a gift bestowed by those who consider themselves knowledgeable upon those whom they consider to know nothing. Projecting an absolute ignorance onto others, a characteristic of the ideology of oppression, negates education and knowledge as process of inquiry*'. From the English edition translated by Myra Bergman Ramos and published by Penguin Education in 1996.

11 '*they "work by themselves"*'. From the English edition translated by Ben Brewster and published by Verso in 2014.

12 '*power affects the body*'. From the English edition translated by Jeanine Herman and published by Semiotext(e) in 1989.

13 '*All the activity of the disciplined individual must be punctuated and sustained by injunctions whose efficacity rests on brevity and clarity; the order does not need to be explained or formulated; it must trigger off the required behaviour and that is enough*'. From the English edition translated by Alan Sheridan and published by Vintage Books in 1995.

14 '*The argumentation required for a scientific statement to be accepted is thus subordinated to a "first" acceptance (…) of the rules defining the allowable means of argumentation*'. From the English edition translated by Geoff Bennington and Brian Massumi and published by the University of Minnesota Press in 1984.

15 '*the appearance in this new art of government of mechanisms with the function of producing, breathing life into, and increasing freedom, of introducing additional freedom through additional control and intervention. That is to say, control is no longer just the necessary counterweight to freedom (…): it becomes its mainspring*'. From the English edition translated by Graham Burchell and published by Palgrave Macmillan in 2008.

16 '*the system of the ideas and representations which dominate the mind of a man or a social group*'. From the English edition translated by Ben Brewster and published by Verso in 2014.

6 Climate change and the ultimate challenge of modernity

Climate change constitutes the ultimate challenge for the project of modernity and its '*unentrinnbaren Zwang zur gesellschaftlichen Herrschaft über die Natur*'[1] (Horkheimer and Adorno, 1947, p. 41). Its global and long-term threat is all that modernity, as envisioned per the Enlightenment, has always endeavoured to control to free people from the dangers of Nature. Climate change is thus a formidable test that the project of modernity is to overcome to ensure the future of humanity (Malm, 2018). This effort is acknowledged by the definitive and universal recognition that is the Nobel Peace Prize:

> By awarding the Nobel Peace Prize for 2007 to the IPCC [Intergovernmental Panel on Climate Change] and Al Gore, the Norwegian Nobel Committee is seeking to contribute to a sharper focus on the processes and decisions that appear to be necessary to protect the world's future climate, and thereby to reduce the threat to the security of mankind. Action is necessary now, before climate change moves beyond man's [sic] control.
>
> (Norwegian Nobel Committee, 2007)

It is therefore no surprise that an entire *dispositif*, in Foucault's terms, has been established to tackle such a challenge of a lifetime or, rather, of humanity's lifespan. This *dispositif* is sustained by a powerful and pervasive discourse and scientific machinery that are enacted by specific policies and actions on all scales, from the global to the household. As for other issues discussed in this book, it is not, by any means, our intention to contest the material reality of climate change and its impact on our physical environment and societies. Climate change is a major issue that deserves due attention wherever its effects are of significance to people's lives and livelihoods.

Rather, our contention in this chapter is that the standalone *dispositif* to address climate change is a mere replication of – and, as such, extends – the broader *dispositif* articulated to address disasters, including its hegemonic and imperialist agenda that overlooks the diversity of local realities around the world. In fact, we will argue that the *dispositif* for climate change takes us back to square one and to the nature/hazard side of the nature/

DOI: 10.4324/9781315752167-6

hazard–culture/vulnerability binary. In this perspective, it shows how resistant and enduring the legacy of the Enlightenment and the associated project of modernity are in addressing hazards and threatening changes in our immediate and global environment. It is ultimately our argument that the dominant Western discourse on climate change actually re-energises the project of modernity at a time when some were promising its demise (e.g. Chakrabarty, 2000; Lyotard, 1979).

In the following sections, we successively focus on the scientific discourse on climate change, with a particular interest in adaptation, and the *dispositif* it supports through the lenses of disaster studies paradigms, as introduced in Chapter 1, and governmentality as the modern art of government, as discussed in Chapter 5. We use a brief example of Kiribati to illustrate how the hegemonic discourse and *dispositif* for climate change are enacted and imposed on the world by the West and its institutional allies in order to sustain their imperialist ambitions. As such, the governmentality of climate change renews the broader disaster risk reduction agenda and provides a new impetus for the project of modernity as thought through the Enlightenment.

Climate change and the resurgence of the hazard paradigm

The study of climate is nothing new and dates back to antiquity. However, never in history has it achieved the prominence it has since the late decades of the twentieth century. Recent significant changes observed in climate patterns across the globe have mobilised tens of thousands of scholars, many gathered in tens of dedicated research institutions and supported by funding worth tens of billions of US dollars. An intergovernmental panel convened by the United Nations, the Intergovernmental Panel on Climate Change (IPCC), has been formed to gather evidence and inform policies. This sophisticated and broad scientific machinery has produced a standalone and solid narrative that supports the contemporary and hegemonic discourse on climate change. This narrative frames a global and indeed very critical issue that affects a myriad of very different cultures and societies within a single and universal discourse.

Moreover, this discourse is informed by an array of concepts, most of which find their origin in Latin etymologies and, as such, reflect Eurocentric and hence Western ontologies and epistemologies. Since climate change is considered primarily a threat or a national security issue (Barnett, 2003, Smith, 2008; the National Security, Military, and Intelligence Panel on Climate Change, 2020) rather than an opportunity, many of these concepts have been borrowed from the broader field of scholarship that focuses on disasters at large. Hence, the discourse on climate change is informed by concepts such as hazard, exposure, vulnerability and resilience, which we already discussed in Chapter 1 (with the exception of exposure, which stems from the Latin *expōnere*). In addition, two specific concepts have emerged in the context of climate change in view of informing policies to address its causes and impacts. These are the concepts of mitigation (from the Latin *mītigāre*) and adaptation (from the Latin *adaptāre*). None of these concepts appeared in Diderot and d'Allembert's *Encyclopédie*, nor were they listed in Chambers's *Cyclopædia.* Yet both are included, with meanings similar to contemporary understandings, in Emile Littré's *Dictionnaire de la langue française*, published between 1863 and 1872, and the *New English Dictionary on Historical Principles* (the ancestor of the *Oxford English Dictionary*), published between 1884 and 1923.

All of these concepts, including mitigation and adaptation, are framed from the perspective of the climate as a threat. It is obvious in the case of hazard, exposure and mitigation. It is also for vulnerability and resilience, whose definitions provided by the IPCC both refer to adaptation in direct relation to nature and its hazards:

> the process of adjustment to actual or expected climate and its effects. In human systems, adaptation seeks to moderate or avoid harm or exploit beneficial opportunities. In some natural systems, human intervention may facilitate adjustment to expected climate and its effects.
> (Intergovernmental Panel on Climate Change, 2014b, p. 118)

This definition is interesting because it takes us back to the work of Gilbert White and other social science pioneers of the dominant hazard paradigm back in the 1930s and 1940s, which we discussed in Chapter 1. It clearly posits that people need to *adjust* to climate change in a skewed dialectical relationship where nature dictates the rules of the game.

Therefore, the two concepts of mitigation and adaptation are meant to capture both sides of the hazard–vulnerability binary and thus continue to entertain the dialectical relationships between nature and culture as the ontological assumption that underpins our understanding of disaster risk in general and those risks associated with climate change in particular. However, the contemporary discourse on climate change and its focus on mitigation and adaptation marks a backward movement of the pendulum, re-emphasising the predominance of nature to the detriment of culture and society.

Unsurprisingly, then, the mitigation–adaptation dialectic appears firmly grounded in the legacy of the Enlightenment and its persistent goal to control nature so that people can flourish in safe societies. As Smith (2008, p. 244) rants, 'it leaves sacrosanct the chasm between nature and society – nature in one corner, society in the other – which is precisely the shibboleth of modern western thought'. In fact, as suggested in the introduction of this chapter, climate change, often depicted through the lens of extreme hazards and dramatic global consequences, offers a 'fantastic' new challenge for the project of modernity. One that has renewed its relevance in a world where nature otherwise had become less of a threat or at least a threat we think we have learnt how to tame. One that has also reinforced the importance of science and technology in understanding and addressing new threats associated with changes in climate pattern *un*-precedented in modern human history as signalled up-front in the IPCC Fifth Assessment Report of 2014 (p. 2): 'warming of the climate system is unequivocal, and since the 1950s, many of the observed changes are unprecedented over decades to millennia'.

As a result, science is concerned with not only furthering knowledge of past and current changes but forecasting upcoming evolutions in climate patterns in a projection towards an *un*-certain future (Demeritt, 2001). It is focused on anticipating this future with new 'normal' temperatures, rainfalls, wind speeds, and so on and potential tipping points beyond which the world may face catastrophic consequences (Castree, 2017). In fact, the importance the scientific narrative and broader discourse on climate change give to the scenarios associated with global mean temperature increase mirrors the significance of the magnitude in defining hazards and that of setting quantitative thresholds to define disasters in general. This quest for pantometry and norms, often related to mean and average values (see Chapter 4), was criticised by Canguilhem (1966) when applied to medicine. Similar concerns can be raised in the context of climate change when tipping points and scenarios that determine the magnitude and significance of the changes in climate patterns are based exclusively on the predictions of science (i.e. Western science) but with very little consideration of local people's views and concerns. Canguilhem (1966, p. 111) actually adds that

> l'importance c'est une notion objective pour le naturaliste, mais c'est au fond une notion subjective en ce sens qu'elle inclut une référence à la vie de l'être vivant, considéré comme apte à qualifier cette même vie selon ce qui la favorise ou l'entrave[2].

If we are to follow Canguilhem's argument, the significance of climate change and the tipping points beyond which its impact becomes damaging and problematic are subjective matters that relate to people's ability to live the lives they wish for themselves and their kin, which depends on a multiplicity of factors other than the climate. This subjective assessment of one's personal condition and aspirations is obviously independent from a statistical deviation from mean or average values of whatever climate indicators/ phenomena/processes.

The predominance of Western science to the detriment of local views and of the subjective representation of climate change and its impact for each and every individual in diverse situations and locations is nonetheless reinforced and widened by the support it gains from international organisations and other donors which fund initiatives that address climate change as well as international non-governmental organisations (NGOs) that implement such actions. These actors contribute to imposing concepts such as vulnerability, resilience, mitigation and adaptation as standards framed from a normative perspective that prevents other views from emerging. In fact, the determination of the IPCC and other international organisations to impose standard definitions for both mitigation and adaptation, through glossaries and other consistent references to the same wording, as well as the wide uptake of these definitions worldwide well reflects the hegemonic, imperialist and normative nature of the discourse on climate change. Bankoff (2019, p. 231) summarises that well:

> [A]daptation is very much a top-down rather than a bottom-up concept largely conceived and implemented by UN and international organisations. Its definition and application are fought over in much the same way as were vulnerability and resilience.

As a result, it seems safe to argue that the scientific narrative on climate change and the broader discourse it supports are in no way new, nor do they provide any ground-breaking pathway towards understanding people's response to hazards in our environments. It is firmly grounded in the legacy of the Enlightenment and its modern *épistémè*. Its significance and traction lie in its ability to re-energise the project of modernity by depicting nature as a re-emerging threat to humanity. One that deserves a standalone and sophisticated global *dispositif* to address the unique nature and dimension of the problem.

On the governmentality of climate change

The positioning of climate change as a unique, *un*-precedented and global threat to humanity, unlike the processes and events commonly identified as disasters, has justified that a standalone *dispositif* be developed. Because the nature of the problem and its ontological and epistemological framing are nonetheless similar to those that underpin disasters, we should not be surprised that the climate change *dispositif* is similar to the one deployed to address disasters. A *dispositif* that sustains a similar form of government that is governmentality.

At the international level, the *dispositif* for climate change is articulated around the United Nations Framework Convention on Climate Change and the two different treaties it encompasses: the 1997 Kyoto Protocol and the 2015 Paris Agreement. These provide regulatory and normative injunctions towards mitigation and adaptation informed by the scientific evidence provided by the IPCC. Injunctions include the obligation (because it is a binding requirement of both the 1997 Kyoto Protocol and 2015 Paris agreement) to reduce emissions of greenhouse gases as well as strong recommendations to conserve sinks and reservoirs for the same gases, foster adaptation, strengthen mechanisms to respond to the impacts of climate change (including those associated with *extreme* hazards), and enhance initiatives towards education and awareness. Most of these injunctions obviously mirror – in fact, often duplicate – strategies designed to reduce disaster risk, strategies that are hazard/climate-driven.

The universal relevance of all these normative injunctions is explicitly justified in the name of human rights and the need to protect the future of humankind, a sort of *raison du monde* that supersedes Foucault's *raison d'état*. As a result, the reporting and accountability mechanisms imposed on signatory governments are stringent. As for disaster risk reduction, the reporting and accountability mechanisms are tied to funding schemes and other political strings associated with global neoliberal policies imposed by Western governments and their allies among international organisations. As such, the United Nations Framework Convention on Climate Change draws on the alleged divide between a safe North and a dangerous South and justifies the leadership of the former by its responsibility for emitting more greenhouse gases, as per the 2015 Paris Agreement:

> Developed country Parties should continue taking the lead by undertaking economy-wide absolute emission reduction targets. Developing country Parties should continue enhancing their mitigation efforts, and are encouraged to move over time towards economy-wide emission reduction or limitation targets in the light of different national circumstances.

In fact, this divide between a responsible but safe North and a 'victim' and vulnerable South constitutes a powerful rationale for justifying the transfer of technology, financial resources, and capacity-building that are the core of the 2015 Paris Agreement. As a result, bilateral donor agencies from Western governments, international organisations, development banks and philanthropic institutions provide the governments of 'vulnerable' countries and NGOs that support the people of these countries with technological resources, funding opportunities and technical and other forms of training.

At the national level, governments are strongly encouraged to conduct their own assessment of the impacts of climate change, based on scenarios of climate patterns provided by (Western) science (Demeritt, 2001). These provide the tangible and usually quantitative evidence of both climate impacts and available assets and resources to support decision-making in an econometric approach to governing that Foucault (2004a) referred to as political economy. These national and local assessments of the actual and future impacts of climate change are meant to be the basis for eventually developing national and local adaptation plans and legislations. These plans are to be implemented by standalone institutions that, in most cases, are organised after a top-down hierarchical structure as for disaster risk reduction. These institutions constitute the tangible architecture of the *dispositif* for climate change. However, the significance of the problem and its positioning as the single most important threat to humanity have led climate change to pervade almost all dimensions of society, including education, health, food, agriculture and forestry, fisheries and industry.

Climate change and its *dispositif* thus provide fertile ground for the disciplinary and regulatory mechanisms of subjection that underpin the modern art of government that is governmentality while it promises liberation from the hazards of nature. It imposes normative behaviours with regard to everyday habits, whether they are associated with water and energy consumption or food and health as well as decisions that may affect people's everyday lives in the long term. These injunctions penetrate the body of individuals when they have to learn to switch off lights, turn off taps, and so on, on a daily basis, so that these small gestures become common sense in an exemplary case of what Foucault called social orthopaedics. The discourse on climate change infiltrates even deeper in bodies when the injunction is tainted with a Malthusian argument and now suggests fertility controls in order to reduce greenhouse gas emissions in the future. As such, people are insidiously subjected to a normative discourse on climate change as framed from the perspective of Western science.

This process of subjection also pervades the injunction to plan for the future and the associated requirement to organise and control space and time, Foucault's *quadrillage* of society. Therefore, actions towards both mitigation and adaptation are designed to maximise spatial resources based on given climate scenarios at specific time frames. An example is through the so-called managed retreat in dealing with sea-level rise. There is, in such strategies of *retreat*, an inherent allegory to military tactics that pervade many other dimensions of climate change mitigation and adaptation. Examples are the 'fortification' of shorelines to 'defend' coastal regions of the world, the development of warning systems, and the 'tactical' planning of land use in facing climate hazards.

Thus, the *dispositif* for climate change relies primarily on technocratic and technological fixes geared towards controlling nature, reminiscent of the hazard paradigm for dealing with disasters that dominated back in the 1970s and 1980s when natural hazards, in their *extreme* dimension, were seen as major threats to economic growth and development (Bankoff, 2001).

Nonetheless, a significant parallel imperative has emerged that emphasises the importance for climate change policies and actions, especially those focused on adaptation, to be inclusive and to foster the participation of those most at risk. This is the so-called 'community-based/led/managed' climate change adaptation. This parallel injunction mirrors the discourse on inclusion and participation in disaster risk reduction which we discuss in Chapter 7. For example, the United Nations Framework Convention on Climate Change encourages the participation and inclusion of those most vulnerable to climate change, with a particular focus on indigenous people, migrants, people with disabilities, children and women, in a narrative that parallels that of the Sendai Framework for Disaster Risk Reduction signed a few months before.

That this inclusion and participation imperative emerged in parallel with the dominant hazard/climate-driven discourse on climate change is largely due to the fact that climate change became a significant global issue at the same time as the broader discourse on inclusion and participation rose to prominence in the 1980s and 1990s and that the two discourses were similarly seized by actors of development such as international organisations and NGOs. Therefore, one may argue that, taken together, as acknowledged in the United Nations Framework Convention on Climate Change, these parallel discourses can be seen as an attempt to bridge the two sides of the nature/hazard–culture/vulnerability binary. A sort of synthesis of decades of learning in the broader fields of development and disaster risk reduction in the historical dialectic of Hegel.

However, one is also to note that the inclusion and participation imperative in climate change adaptation often overlooks the underpinning unequal power relations that prevent access to resources that foster adaptation and thus replicates most of the shortcomings that we discuss in Chapter 7 (e.g. Delica-Willison et al., 2017; Few et al., 2007). As such, the injunction for inclusion and participation in adaptation is a mere repackaging of the normative and regulatory nature of the governmentality of climate change. One that reflects a rebranding from a normative top-down approach, to one that is normative from the 'bottom up'. In fact, the ontological emphasis that the dominant discourse on climate change places on nature as a threat and future climate as a long-term challenge often clashes with local people's concerns, especially beyond the West.

The future is now in Kiribati

There is indeed a significant gap between the dominant discourse on climate change on the international scene and the everyday concerns of local people who deal with climate change. This is particularly true in those places which are likely to be the most threatened by sea-level rise and the other possible harmful effects of changing climate patterns. As a result, despite political commitment and significant institutional support (including funding), programmes to foster adaptation, in particular, lead to very few tangible outcomes at the local level. This gap between international and local priorities is evident for the low-lying atoll countries of the Pacific.

The government of Kiribati has been very vocal regarding the potential impact of climate change for this small island country located in the midst of the Pacific Ocean (e.g. Helvarg, 2010; Leoni, 2012). The potential disappearance of the whole country because of sea-level rise is the ultimate argument used to attract attention to this nation which has long been neglected in international policies.

In the 2000 and early 2010s, the awareness campaign concerning the fate of Kiribati was quite successful. The World Bank and Western government partners awarded the country a series of multi-million-dollar grants under the Kiribati Adaptation Project (KAP). KAP was designed by international experts who emphasised the need for 'improving the protection of public assets' in facing rising seawaters through the reinforcement of existing infrastructure, notably the road network, the construction of seawalls, and the planting of mangrove trees. It also aimed at improving the management of the atolls' scarce groundwater resources (Donner and Webber, 2014;

Kiribati Adaptation Project, 2011; Storey and Hunter, 2010). In addition, the government negotiated with its counterpart in Fiji, who had already welcomed people from Banaba, after the islet became a large phosphate mine, to buy land on Viti Levu. Although this acquisition was officially meant to provide opportunities for economic development and ensure food security for Kiribati, it has been interpreted by many, including i-Kiribati people, as an early move in preparation for the possible progressive relocation of some of the atoll country's population (Hermann and Kempf, 2017).

Despite such governmental lobbying, increasing international attention and their own observations of changes in climate patterns, people deliberately choose to focus on sustaining their daily needs in the context of raising prices of commodities (Zuñiga, 2007). Those everyday needs which matter most to people include purchasing food stuff, paying for school fees, bills and taxes, and securing health care, all of which are largely dependent upon access to stable employment, reliable inter-island transportation, and traditional, social, and cultural values (Kuruppu, 2009; Thomas and Tonganibeia, 2007; Watters, 2008; Zuñiga, 2007). Such pressing daily issues do not preclude local people from noticing modifications in their environment associated with climate change but these are of lesser importance if they do not impact people's ability to make a decent living in the short term (Kuruppu and Liverman, 2011; MacKenzie, 2004).

Many studies have long emphasised the predominance of short-term livelihood strategies in i-Kiribati culture (e.g. Lundsgaarde, 1967; Thomas, 2001, 2002; Watters, 2008). Those first and foremost include fishing and farming complemented by remittances from relatives who have settled in the capital atoll of Tarawa or overseas. For i-Kiribati people, anticipating and planning for the future have not been prominent strategies. Planning on the household scale has recently emerged as the result of new needs brought along Western standards, such as paying for school fees and government taxes (Asian Development Bank, 2002; Watters, 2008). Nonetheless, over the past 1900 years, traditional short-term strategies have not stopped the population of Kiribati from growing dramatically despite continuously facing serious environmental and social challenges such as severe droughts, near famines, introduced diseases, and significant changes in climatic patterns (Lundsgaarde, 1966; Nunn, 2007; Rainbird, 2004). In fact, the i-Kiribati society has proven highly adaptive. As Knudson (1981, p. 97) put it:

> the Gilbertese have adapted their society to new and changing environmental circumstances, and in doing so Gilbertese society has evolved to a new and different form. But it still carries elements that appear to be unchanged; furthermore, these unchanged elements continue to be adaptive in quite different circumstances.

National and local institutions are similarly uncomfortable with the idea of planning as imposed by the former British colonial government and contemporary international aid agencies (Thomas and Tonganibeia, 2007;

Watters, 2008). The successive national development plans have largely remained promises on paper which have yielded limited outcomes on the ground (Asian Development Bank, 2009). In 2012, according to senior staff of the Ministry of International and Social Affairs, none of the island councils had yet effectively implemented the Local Government Act of 2006, which provides guidelines for local development planning, including an item for 'drought and famine relief' (Republic of Kiribati, 2006). Likewise, national authorities acknowledged that the National Disaster Risk Management Plan, which covers issues associated with climate change, crafted in 2010 under the leadership of the scientific division of the Pacific Community (Republic of Kiribati, 2010), had yet to materialise in concrete actions at the local (island) level.

In Kiribati, in contrast to the Western normative, disciplinary and regulatory governmentality, the traditional art of government relies on local meetings held in the communal house named *maneaba* (Maude, 1980). It draws upon 'the political concept that people must meet together to decide on their own welfare and solve problems' (Tabokai, 1993, p. 24). Although the traditional authority lies within the hands of elders, the *maneaba* political system emphasises economic and social equality between local people who are encouraged to contribute and share food (Lundsgaarde, 1966; Tabokai, 1993).

In this situation, Western-led projects, be they for climate change adaptation or development at large, are most often considered by local leaders for the short-term benefits (that provide potential incomes) rather than their long-term outcomes (Mallin, 2018). Each project intended to be implemented in any of the atolls of Kiribati has to be approved by the local island council, which, in exchange, requires that a significant sum of money be distributed to every member of the council. This practice, which should not be misinterpreted as what the West calls corruption, is actually institutionalised by the Ministry of Internal and Social Affairs. Similarly, activities organised at the local level by outsiders under a so-called 'participatory' approach most often entail that each participant in a group discussions and workshops receive financial compensation.

The gap between the objectives and expectations of foreign-designed projects, and how local institutions and people perceive them in the context of pressing everyday needs and day-to-day response, is obvious. Projects are viewed as potential resources not for their intended long-term outcomes but for the opportunity they offer to meet daily needs and sustain other social functions, as, in the past, Lundsgaarde (1967, p. 25) observed that 'the introduction of money has been adjusted to conform to the traditional system of social reciprocity and obligation and without serving all the economic functions of money in Western society'. Over the past five decades, the monetisation of the economy has made financial incomes essential to sustain daily needs in Kiribati (Kuruppu, 2009; Watters, 2008; Zuñiga, 2007). At the local level, project funds therefore are siphoned for short-term purposes.

In that context, what is the impact of climate change–related projects in Kiribati? The first two phases of KAP (KAP I and KAP II) led to effective but very limited concrete outcomes (Donner and Webber, 2014), which include four seawalls and the reinforcement of two causeways, a set of infiltration galleries for a local school, a few rainwater harvesting devices, the rehabilitation of some water tanks, a small number of mangrove trees, and finally a series of studies to assess the state of groundwater reserves. In addition, these activities first and foremost focused on Tarawa with limited benefits for the outer islands. These were useful but limited tangible outcomes for a project which had, as of 2012, required more than $6 million (USD).

The KAP mid-term review conducted in 2008 by foreign consultants hired by international donor agencies acknowledged the shortcomings of the programme. However, it failed to recognise the root causes of the problem and the unsuitability of long-term actions in the context of Kiribati. Rather, the review pinpointed people for their inadequate behaviour and recommended 'behaviour changes' – or social orthopaedics, in Foucault's vocabulary – on the side of local people. As a result, a series of awareness-raising and behaviour change campaigns were designed, again by foreign consultants largely unaware of local realities (Menzies, 2009; the World Bank, 2010).

In Kiribati, as in many other places in the world, the international discourse on climate and its lingering projects at the national and local levels are disconnected from local realities, including people's needs, the cultural fabric and the traditional art of government. In a brilliant analysis, Webber (2013) goes further and argues that vulnerability to climate change is actually performed, in Butler's sense, in order to secure funding so that

> some donors require i-Kiribati bureaucrats to marshal their vulnerability to climate change in order to be successful in their applications for financing. Performances, funding regimes, and materialities are reframing Kiribati and its development assistance regime in terms of vulnerability to climate change. This framing creates climate vulnerability as the overriding concern, minimizing other development and local interests, in such a way as to make Kiribati once more the subject of international and transnational forces, institutions, and interests. Climate change adaptation is a paradigm, a discursive frame, and a metatrope of development assistance, which is shaping the way international assistance is distributed and making possible the conditions for new performance and practices of vulnerability.
>
> (Webber, 2013, p. 2725)

Obviously, the aggressive discourse of the government in dealing with climate change on the international scene has proved successful in helping foreign stakeholders locate Kiribati on the world map and also in attracting a good deal of attention from funding agencies. Yet it has largely failed to consider the daily priorities and needs of local people whose wellbeing has not been lifted over the past few decades (Asian Development Bank,

2009; Storey and Hunter, 2010; Zuñiga, 2007). In fact, some representatives of NGOs and Webber (2013) fear that the realignment of government expenditures towards climate change adaptation projects, and the siphoning of international aid for the same purpose, could lead to significant cuts in other budget allocations, especially for health and education, which are other priorities of the government and which matter most in the short term.

Putting an overarching emphasis on long-term climate change may be considered a harmful distraction from local priorities which are short-term and related to daily life. As we noted elsewhere, 'focusing on a single climate change challenge is dishonest in failing to acknowledge other equally important concerns' (Kelman and Gaillard, 2010, p. 31). In many instances in the Pacific, policies designed to face the long-term impact of climate change have conflicted with day-to-day traditional responses to local concerns, thus eroding indigenous mechanisms to cope with environmental issues, including climate change (Campbell, 1990, 2006).

The inadequate discourse on climate change in the context of local realities is puzzling and worrying since a similar gap was emphasised four decades ago in disaster risk reduction, including in facing climate hazards (Copans, 1975; Hewitt, 1983; Torry, 1979b; Waddell, 1977; Wisner et al., 1977). Unfortunately, lessons have not been learnt, as contemporary policies geared to enhancing the ability of local people to face climate change reproduce the same technocratic measures which have long failed to prevent the occurrence of disasters, as we define them in the West.

Emphasising the importance of everyday needs, local cultures and traditional forms of government does not undermine the significance of climate change and its impact. Rather, it better embeds climate change–related actions within the daily issues which matter most for local communities and institutions. Strengthening people's everyday lives ultimately results in gains for facing climate change, as stronger livelihoods mean lessened vulnerability and greater ability to adapt.

Climate change and Western imperialism

The normative and regulatory approach to climate change observed in Kiribati and elsewhere in the world directly results from the monolithic nature of climate change science as gathered by a single overarching institution that is the IPCC. Since there is only one authoritative voice when it comes to providing advice and recommendations with regard to climate change, it is unsurprising that policy and actions it informs are standardised. As Demeritt (2001, p. 308) pointed out:

> the substance of scientific consensus on global climate change is not as important as the fact that agreements among international community of scientific investigators has enabled them to enrol governments around the world to binding GHG [greenhouse gas] emission reductions.

This monolithic and authoritative scientific narrative is reinforced by the quantitative and normalised, in Canguilhem's terms, nature of the evidence and the alleged absolute truth they offer, truth legitimised by dominant positivist epistemologies and the contemporary hegemonic *métarecit* of the West that is neoliberalism. As such, the contemporary discourse on climate change prolongs and actually reinforces the dominant discourses on disasters, in their alleged diversity, and their hegemonic influence on policies and actions geared towards addressing threats in people's everyday lives, especially those linked to our physical environment.

Again, we do not challenge the material reality of climate change nor the significance of its impacts for those who deal with them on an everyday basis. Our concerns are that a single scientific tradition, which is that of the West as informed by the legacy of the Enlightenment, shapes one single discourse on climate change and that, henceforth, alternative and non-Western interpretations of and responses to changes in climate patterns are unheard or neglected. This is of particular significance when this dominant discourse emerges from past colonial powers and their allies and thus prolongs a colonial legacy that reinforces and legitimates unequal power relationships between a few countries, most of which have a common European heritage, and the rest of the world. Hence, one may agree with Gramsci (1930, p. 12) that '*ancora una volta la "scienza" era rivolta a schiacciare i miseri e gli sfruttati*'[3].

This, once again, raises the question of re-presentation and representation that we posed in Chapter 1. Do the voices from small island countries, such as that of the former president of Kiribati, invited to speak at international conferences in support of a discourse that portrays their islands as disappearing, actually reflect the local, everyday and diverse realities of all people of these countries? Or rather do they justify external intervention through a normative and regulatory approach to mitigating and adapting to climate change? Is the contribution of the numerous scholars from less affluent and non-Western countries to the IPCC useful when these researchers have (consciously or not) to fit within and conform to Western ontologies and epistemologies in studying and addressing climate change? Or are these non-Western voices just a token concession on the part of the West to secure global consent, in Gramsci's terms, to the governmentality of climate change and the broader project of modernity and its alleged universal relevance?

In fact, the polarising divide that the discourse on climate change has fabricated between a polluting (and hence responsible) but safe West (and other powerful countries such as China and Japan) and a 'victim' and vulnerable rest of the world, evident in the 2015 Paris Agreement, contributes to a discourse of alterity that validates and justifies Western science and imperialist intervention to 'rescue' the rest of the world (Bankoff, 2001, 2019). As such, the discourse and governmentality of climate change not only build upon the North–South binary but reinforce it in the name of the universal project of modernity and its scientific, technology and human rights imperatives. More than 40 years ago, Foucault (1978, p. 12) wrote a powerful charge against such imperialism:

> The third and last is the movement by which, at the end of the colonial era, people began to ask the West what rights its culture, its science, its social organization and finally its rationality itself could have to laying claim to a universal validity: is it not a mirage tied to an economic domination and a political hegemony? Two centuries later the Enlightenment returns: but not at all as a way for the West to become conscious of its actual possibilities and freedoms to which it can have access, but as a way to question the limits and powers it has abused. Reason – the despotic Enlightenment.

The governmentality of climate change thus appears as another instrument of imperialism for the West. This instrument of imperialism builds upon the material reality of climate change and of its impacts to impose a normative and regulatory agenda on the rest of the world.

This agenda is inherently associated with a liberal ideology which prolongs the goal of the project of modernity, as envisioned in the eighteenth century, to free people from the dangers of Nature so that they can flourish in life while insidiously fostering discipline and control. In fact, the irruption of climate change into the political arena in the late 1980s not only re-energised the project of modernity by providing it with a new global threat but also coincided with the resurgence of and sustained the relevance of liberalism, as a universal ideology, under the latest iteration that is neoliberalism. The *dispositif* for climate change here appears very powerful through its *quadrillage* of society and its ability (especially when it comes to climate change adaptation) to penetrate people's bodies and plant neoliberal ideas in the minds of individuals who trust that they are saving the world, which they may indeed contribute to, but under the guise of neoliberalism.

In fact, the strong emphasis that the discourse on climate change puts on people's behaviours and the normative injunction to change 'inappropriate' behaviours to both mitigate and adapt to climate change shift some of the responsibility for securing the future of the planet onto individuals to the detriment of the state, whose duty is largely limited to encouraging and monitoring the emission of greenhouse gases. This is where the integration of 'community-based/led/managed' climate change adaptation, at least in its normative iteration (Chapter 7), makes sense as it provides a venue for widely spreading and creating a collective momentum for the inherent neoliberal nature of the governmentality of climate change (Bankoff, 2019).

The *dispositif* for climate change is therefore a very powerful avenue for advancing the neoliberal agenda and its imperialist ambition around the world (Felli and Castree, 2012; Smith, 2008). The global threat to humanity that is climate change constitutes a powerful message that is hardly challengeable, a new *raison du monde*. Since the West is portrayed as both the legitimate leader because of its responsibility for the occurrence of climate change and the only solution to the problem through its science and technology, the Western art of government that is governmentality appears as the ultimate way forward. As such, it appears relatively easy to make the Western discourse on climate change pervade all dimensions of society, as briefly discussed earlier, through multiple channels, including the media, cultural products (whether they are films, novels or video games), schools and many others. The multisource, constant and repetitive nature of the injunctions that these actors and vectors of the *dispositif* for climate change convey to virtually all members of society, in one way or another, contributes to making the Western discourse common sense, including in places of the world where this discourse does not fit within local cultures and world views/senses. In this perspective, it reproduces what Fanon (1952, p. 124) so aptly captured in *Peau noire, masques blancs*:

> (…) il y a une constellation de données, une série de propositions qui, lentement, sournoisement, à la faveur des écrits des journaux, de l'éducation, des livres scolaires, des affiches, du cinéma, de la radio, pénètrent un individu – en constituant la vision du monde de la collectivité à laquelle il appartient. Aux Antilles cette vision du monde est blanche parce que qu'aucune expression noire n'existe[4].

The picture portrayed here is one of hegemony, in Gramsci's sense. The Western discourse on climate change is legitimised by the alleged irrefutable evidence of Western science, and the portrayal of the threat as one of unique significance for the survival of humanity. This discourse is shared constantly and universally through intellectuals such as scientists, the media and popular cultural figures, so that it become common sense. As Spivak (1993b, p. 144) once said, 'the devout colonial subject, decent dupe of universalism, thinks to learn the trick perfectly'. In the end, this process legitimises the disciplinary and regulatory modern art of government of the West that is governmentality and its contemporary ideology that is neoliberalism because the West, its science and technology appear as the only available opportunity to address climate change and secure the future of humanity. This seemingly irrefutable reality contributes, in the end, to securing nearly universal consent from all sectors of almost all societies around the world. As Quijano (2000a, p. 222) concludes in speaking on behalf of the people of Latin America:

> the tragedy here is that we all have been led, knowingly or not, willingly or not, to see and to accept that image [Eurocentric] image as our own reality and ours only. Because of it, for a very long time we have been what we are not, what we never should have been and what we never will be. And because of it, we can never catch our real problems, much less solve them, except in only partial and distorted way.

Climate change and the demise of the project of modernity?

Re-energising Western modernity as a project of universal relevance takes us back to its roots and the clear demarcation established between nature and culture. Climate change indeed provides an opportunity to once again picture nature as a threat to humanity. A hazard that needs to be controlled if the aspiration of the Enlightenment to offer people with a chance to flourish in life is to survive the twentieth century. It had indeed become obvious that the project of modernity was running out of steam. Nature had become increasingly commodified so that its ontological polarity with culture, foundational to the project of modernity, was not so evident anymore, with the exception of *extra*-ordinary and *un*-expected natural hazards that lead to sporadic disasters. Consequently, the critiques of modernity, in the path paved by Horkheimer and Adorno (1947), had become more vocal and had increased in numbers across all dimensions of society, be they the arts and popular culture (Foster, 1983), architecture and planning (Harvey, 1990), and the sciences (Feyerabend, 1975, 1987; Latour and Woolgar, 1979). The ascent of postmodern thoughts to prominence in the 1970s and 1980s marked this critical turn, which could have signalled the demise of the project of modernity.

Climate change thus becomes handy for the West because such a reaffirmation of the universal relevance of the project of modernity intrinsically refuels its imperialist and neoliberal agenda and, as such, prolongs the legacy of colonialism in a process akin to Quijano's (1992) coloniality of power. Coloniality, in Quijano's (2000b, p. 381) words, refers to

> una estructura de dominación/explotación, donde el control de la autoridad política, de los recursos de producción y del trabajo de una población determinada lo detenta otra de diferente identidad, y cuyas sedes centrales están, además, en otra jurisdicción territorial[5].

Indeed, as Quijano (2000a) and Mignolo (2011) have shown, modernity and coloniality are organically linked so that doping one inherently boosts the other. Not only has colonialism supported the growth of Europe through the extraction of natural resources and staff power, it has also created a binary view of the world that casts one half of it as 'retarded' or 'traditional' (what we nowadays call the South) and one that is 'advanced' and 'developed' (i.e. Europe and its off-shore settlements). This picture of alterity and unequal ability to 'progress', which emerged throughout the sixteenth and seventeenth centuries, has contributed to legitimising the leadership of Europe and, a century later, to that of the project of modernity and its governmentality henceforth projected as of universal relevance. Escobar (1995) and, in the specific context of disasters, Bankoff (2001, 2003a, 2019) have unpacked how the Western project of modernity, since then, has ambitioned to 'develop' and 'save' the rest of the world from the wrath of diseases, poverty, natural hazards and terrorism through its science and technology. Since its emergence as a global issue in the 1980s, climate change has been portrayed in this perspective, which further divides the world under a seemingly universal problem, re-energises the project of modernity that started to run out of steam, and finally fuels the neoliberal and imperialist agenda of the West.

In conclusion, our argument is, once more, not to dismiss the significance and relevance of climate change as a global challenge. Neither do we aim at dismissing the relevance of Western science wherever it makes sense and fits within local world views. Our contention is that Western science and the governmentality it sustains should not be rolled out around the world as part of an imperialist agenda which aims to subject the rest of the world to the sole views of the West, including its understanding of climate change and how to respond to it. Rather, we argue that climate change should be understood from multiple perspectives that reflect local everyday realities and diverse world views/senses. In fact, if Western science is to draw upon the principle of reason, as its epistemological foundation suggests, one may agree with Quijano (1992) that there is nothing less reasonable than to assume that one single understanding of the world, and of climate change in particular, does actually exists and that hence there is one unique way of dealing with it as the normative governmentality of climate change proposes.

Notes

1 '*inescapable compulsion toward the social control of Nature*'. From the English edition (p. 27) translated by Edmund Jephcott and published by Stanford University Press in 2002.
2 '*For the naturalist importance is an objective idea, but it is essentially a subjective one in the sense that it includes a reference to the life of a living being, considered fit to qualify this same life according to what helps or hinders it*'. From the English edition translated by Carolyn R. Fawcett and published by Zone Books in 1991.
3 '*Once again, "science" was used to crush the wretched and exploited*'. From the English edition translated by Quintin Hoare and published by Lawrence & Wishart in 1977.
4 '*there is a constellation of postulates, a series of propositions that slowly and surely - with the help of books, newspapers, schools and their texts, advertisements, films, radio - work their way into one's mind and shape one's view of the world of the group to which one belongs.15 In the Antilles that view of the world is white*'. From the English edition translated by Charles Lam Markmann and published by Pluto Press in 1986.
5 '*a structure of domination/exploitation, where the control of the political authority, means of production, and labour of a given population is exercised by another population who claims a different identity and whose government is located in another territorial jurisdiction*'. Our translation.

7 Exclusive inclusion and the imperative of participation

Disaster risk reduction these days has to be inclusive. As per the Sendai Framework for Disaster Risk Reduction (United Nations International Strategy for Disaster Reduction, 2015):

> Disaster risk reduction requires an all-of-society engagement and partnership. It also requires empowerment and inclusive, accessible and non-discriminatory participation, paying special attention to people disproportionately affected by disasters, especially the poorest. A gender, age, disability and cultural perspective should be integrated in all policies and practices, and women and youth leadership should be promoted. In this context, special attention should be paid to the improvement of organized voluntary work of citizens.

What reads here as an imperative actually formalises a widespread trend observed over the past four decades. Fostering the inclusion of those who suffer most from what we call disasters has indeed become a core tenet of the actions of many, if not most, non-governmental organisations (NGOs). These actions have been supported by funding from a range of organisations, comprising international organisations and government donor agencies for which inclusion has also become a priority.

The main rationale behind the injunction for inclusion is well known and has been backed by a very broad range of so-called 'evidence' from both research and practice, dating back to the 1970s (Delica, 1993; Heijmans, 2009; Maskrey, 1984, 1989) at least. It draws on three main threads of argument:

1. The local dimension of disasters is crucial. The people who are affected are also the first line of defence and the first responders, whether in dealing with fast-onset or protracted hazards.
2. These people are not 'helpless vulnerable victims'. They display a range of capacities and their behaviour in facing the immediate impact of the disaster is rational and appropriate in the vast majority of cases.
3. These people are also the most interested in reducing risks in their diversity because it is their lives and livelihoods that are at stake. In addition, no one understands the local context better than the locals.

DOI: 10.4324/9781315752167-7

These are all very legitimate concerns and reasons that we fully support in principle. No misunderstanding. Things become a bit more complex in practice though. The agenda towards inclusion, as per the above quote, rightly emphasises the central role of empowerment as a concept, a process and an outcome. By definition, empowerment entails a genuine sharing of power for the benefit of those who are encouraged to participate and who are the most at risk in the dominant narrative on inclusion (Rowlands, 1995). It is a process 'by which people, organizations, and communities gain mastery over their lives' (Rappaport 1984, p. 3).

Fostering inclusion is therefore intrinsically a political agenda. A tall order in the context of the overall governmentality of disaster risk reduction (discussed in Chapter 5) and an even taller order when considering the existing distribution of power across and within many societies in the world. We contend in this chapter, that this political agenda is hardly or at best only partly achieved. Indeed, our concern is that the discourse on inclusion and its international imperative in policy and practice has been de-politicised and de-contextualised. As a result, it has become a new hegemonic *métarécit* that perpetuates Eurocentric/Western and normative framings of disaster and disaster risk reduction and contributes to their diffusion outside of the West.

On the ethos and principles of inclusion in disaster risk reduction

Fostering inclusion in disaster risk reduction, as across all other dimensions of everyday life, emerged and was conceptualised as a political agenda back in the 1980s (Delica, 1993; Maskrey, 1984, 1989). In a seminal paper informed by the work of Engels, Maskrey (1984, p. 28) summarised the agenda, then called community-based hazard mitigation, as follows:

> Community based mitigation, in contrast with its government based counterpart, should not be considered as a 'categorical' program, with the sole objective of mitigating the risk faced by given vulnerable elements to a given hazard in a given moment. Rather, it should be defined as a process of transformation of the social relations of production in a given society, resulting in the progressive reduction of the vulnerability, through the specific paradigm of hazard mitigation. In other words, it would seem to have two different, but at the same time, interrelated objectives:
>
> • progressively reduce the vulnerability of the poor majority of the population, by means of the transformation of the social relation of production;
> • minimize the suffering of the poor majority, during this process, by means of the mitigation of the risk faced by those elements actually vulnerable to a given hazard in a given moment.

The inclusion-through-participation agenda then extended the movements for the liberation of those people who were oppressed by colonial or totalitarian states, such as those initiated by Aimé Césaire and Frantz Fanon in the Caribbean and Africa and Paulo Freire in Latin America. It was then conceptualised not only as a form of resistance to the dominant hazard-driven discourse on disasters but more broadly as a hook for addressing broader political, social and economic issues. The suffering of those affected by disaster was a powerful argument to justify a fairer distribution of resources and opportunities within society.

Indeed, fostering inclusion in disaster risk reduction is all about sharing power and the recognition of others and otherness (Cornwall, 2004; Gaventa, 2006; Hore et al., 2020; Kothari, 2001). It is about facilitating the participation of those most at risk in decisions that affect their everyday lives, in recognition that everyone has unique concerns. As such, inclusion and participation are both processes rather than a means or an outcome. Inclusion requires addressing unequal distributions of power and resources within society, a process that may conflict with local cultural norms and values. Encouraging and actually sharing power for the benefit of those who are usually excluded from decisions are therefore complex, sensitive and often conflictual. It requires constantly questioning who is making decisions, on whose behalf, and based on whose knowledge and ultimately to the benefit of whom (Chambers, 1997; Maskrey, 1984). Making sure that these who/whose/whom questions centre on the same people is at the core of the inclusion-through-participation agenda. Accountability thus is oriented inherently towards those who claim inclusion or are to be included.

The practice of inclusion through participation in disaster risk reduction has been informed by the strategies, techniques and toolkits used to encourage the broader participation of the poor in decisions that affect their daily lives. Back in the 1980s and early 1990s, these strategies, techniques and tools were gathered in what was then called Participatory Rural Appraisal (PRA). Later in the 1990s and continuing since then, PRA has been known under the broader appellate of Participatory Learning and Action (Chapter 4). The shift from PRA to Participatory Learning and Action reflected not only a broadening of scope, from rural to all contexts, but more importantly a recentring of its ethos on learning and sharing to the detriment of the mere appraisal of concerns and resources (Chambers, 1994, 1997). In this perspective, Chambers (1997) has provided a powerful framework for approaching inclusion through participation in practice.

Chambers's framework relies on three tenets that are the tools used to foster participation, the attitudes and behaviours of all involved parties, and ultimately the sharing that both the tools and attitudes/behaviours allow. These three tenets are intrinsically imbricated and it is only the combination of all three taken as a whole that leads to genuine inclusion through participation, which is a process that fosters the sharing of power. Furthermore, these three tenets must be taken as loose directions that do not include a strict list of tools to be used or attitudes to follow in any context. Because it is intrinsically driven by the local context and by local people, the process of inclusion through participation has to be locally grounded and embedded within local priorities, practices, values and norms. As such, there cannot be any silver-bullet approach that applies universally, as Maskrey (1984, pp. 30–31) warned us almost 40 years ago:

> if community-based hazard mitigation is a process which has its roots in the community itself, and is not a categorical program, which happens to be carried out at the community level, then it should be obvious that it cannot he subjected to standard 'recipes' for its implementation. Local cultural, social, economic, political and physical conditions are infinitely variable and form extremely complex backcloth, in front of which community based mitigation may or may not take place in any given context. Any analysis of the potential for community based mitigation must take as a starting point the evaluation of these local conditions and not based itself on preconceived solutions. Cases of solutions looking for problems to solve, rather than analysis of problems giving way to specific solutions, are all too common in hazard mitigation.

In addition, Chambers's framework and Maskrey's recommendations have to be taken holistically, across all dimensions of everyday life and society. It does not apply only to the facilitation of one-off, short-term initiatives. As Chambers (1983, pp. 168–169) continues:

> for those who are poor, physically weak, isolated, vulnerable, and powerless, to lose less and gain more requires that processes which deprive them and which maintain their deprivation be slowed, halted and turned back. These reversals have many dimensions. (...) But three dimensions deserve special attention because they combine potential impact with feasibility. They concern reversals in space, in professional values and in specialisation.

Both Chambers's foundational principles and Maskrey's pioneering agenda for disaster risk reduction embrace the political nature of the inclusion agenda. It is organically focused on 'putting the last first' and 'putting the first last' (Chambers, 1983, 1997). An agenda whose 'aim is to transfer more and more power and control to the poor' (Chambers, 1983, p. 147) in 'the objective of transforming society' (Maskrey, 1984, p. 28).

In the specific context of disaster risk reduction, fostering inclusion through participation contributes to both harnessing people's capacities and reducing their so-called vulnerability. The former requires an endogenous process that builds upon local people's intrinsic resources, knowledge and skills. It is often the building block of inclusion endeavours because it is less dependent on power relations with outsiders. On the other hand, reducing vulnerability, as defined in previous chapters of this book, requires that those with power grant access to means of protection in dealing with natural hazards as well as the power to make decisions to those who do not have the same privilege.

This process is, by nature, all about power relations and requires time and transformation of the broader distribution of resources within society. As such, fostering inclusion through participation of the most vulnerable cannot happen in a silo and has to be considered within broader relationships of power on multiple scales, from the local to the global, and across these scales. As Maskrey (1984, p. 29) suggested in his agenda-setting paper:

> community based hazard mitigation does not signify the implementation of mitigation measures exclusively at the community level of action, but rather the progressive activation of all the levels, starting with the community itself. The key issue is "who decides".

In order to achieve such transformation of power structures within society, the process of inclusion through participation has to foster a continuing dialogue between those who relinquish power and those who gain power, which Chambers's (1997) framework captures through both the process of sharing and attitudes and behaviours. This dialogue, according to Freire (1968, IV, p. 20), is absolutely essential

> por isto é que, sendo a ação libertadora dialógica em si, não pode ser o diálogo um a posteriori seu, mas um concomitante dela. Mas, como os homens estarão sempre libertando-se, o diálogo se torna uma permanente da ação libertadora[1].

In practice, though, the dialogical and political nature of the inclusion agenda has largely vanished.

From theory to the practice of exclusive... inclusion

It is not our intention to provide in this section, nor in the subsequent ones, an extensive critique of both the discourse and practice of inclusion and participation in disaster risk reduction. Many excellent reviews of the kind are available in the literature on disasters (e.g. Cannon, 2014; Maskrey, 2011; Titz et al., 2018) and, more broadly, in development studies (Cleaver, 1999; Cooke and Kothari, 2001; Mansuri and Rao, 2013; White, 1996; Williams, 2004). Rather, we selectively focus on some key issues that reflect how this discourse and practice of inclusion through participation are skewed by the hegemony of Western ontologies and epistemologies in understanding and addressing what we usually call disasters and, as a result, how both this discourse and practice perpetuate the imperialist agenda of the West. As such, it is not the ethos and principles of inclusion per se that we critique hereafter, but the proclamation of some of these as universal and their implementation in a blanket fashion that seldom considers the very diverse local realities of the contexts where they are rolled out.

First and foremost, it is essential to distinguish initiatives driven by the very people who claim inclusion in disaster risk reduction from those led by outsiders, in all their diversity. Mansuri and Rao (2013, pp. 31–32) write:

> Organic participation is usually driven by social movements aimed at confronting powerful individuals and institutions within industries and government and improving the functioning of these spheres through a process of conflict, confrontation, and accommodation. (…) Induced participation, by contrast, refers to participation promoted through policy actions of the state and implemented by bureaucracies (the "state" can include external governments working through bilateral and multilateral agencies, which usually operate with the consent of the sovereign state).

As Mansuri and Rao (2013) observe, organic participation is often successful because it is grounded and builds upon local political momentum and leadership. There are initiatives of this kind in disaster risk reduction that need to be recognised and applauded (e.g. Abinales, 2012; Bajek et al., 2008; Bhatt, 2017; Carlton and Mills, 2017; Delica, 1999; Kelman and Karnes, 2007; Kenney and Phibbs, 2015; Luna, 2003). Most often, though, inclusion is induced by outsiders, including international organisations, donor agencies, NGOs and (increasingly nowadays) the private sector. It is this approach that we endeavour to scrutinise in the subsequent sections of this chapter.

Indeed, inclusion has been at the core of the disaster risk reduction and broader development agenda since the 1990s. It is an imperative dictated by the terms of international agreements specific to disaster risk reduction, such as the Sendai Framework for Disaster Risk Reduction, as well as broader goals such as those listed in the Universal Declaration of Human Rights. It is further fuelled by the neoliberal agenda that pervades many dimensions of international and national government policies, which put a growing burden on the individual to the detriment of the state (Cornwall and Brock, 2005; Leal, 2007).

The whole discourse on inclusion in disaster risk reduction, as in broader society, is based on the assumption that those who are meant to be included stand outside of society, or at its margin, in the first place. Hence, the need to pull them in or towards the centre – that is, to literally include them. This assumption is fundamental because it entails a process of integration or recentring, a process that is mechanical rather than political. As Freire (1968, II, p. 5) brilliantly exposed:

> Na verdade, porém, os chamados marginalizados, que são os oprimidos, jamais estiveram fora de. Sempre estiveram dentro de. Dentro da estrutura que os transforma em "seres para outro". Sua solução, pois, não está em "integrar-se", em "incorporar-se" a esta estrutura que os oprime, mas em transformá-la para que possam fazer-se "seres para si?".[2]

This dialectic of the inside and outside, of the centre and the margin, has been a powerful one. It casts society in halves, reinforcing a binary view of the world. It mirrors the divide of the world into a safe North and danger-ous South and of knowledgeable experts versus ignorant locals, which we discussed in previous chapters. This divide further justifies a practice of inclusion that draws on external 'intervention' rather than organic mobi-lisation. This is Perlman's (1976, p. 247) argument in her landmark book *The Myth of Marginality*:

> Ironically, the myth of marginality is itself a real material force – an ide-ology which informs the practice of the dominant classes and has deep historic roots in the history of Latin American cities. It is a vehicle for interpreting the social reality in a form which serves the social interests of those in power. A myth is merely a strongly organized and wide-spread ideology which, to use Karl Mannheim's definition, develops from the "collective unconscious" of a group or class and is rooted in a class-based interest in maintaining the status quo. It involves a belief system, a systematic distortion of reality reflected in this system, and a specific function for those ideas in serving the interests of a specific group.

As discussed earlier in this chapter as well as in Chapter 5, this dialectical discourse on inclusion has pervaded almost all dimensions of the *disposi-tif* of disaster risk reduction. The proclamation of inclusion as a universal imperative, as per the recommendations of international agreements such as the Sendai Framework for Disaster Risk Reduction, is the latest expres-sion of the normative and regulatory nature of disaster risk reduction, one that seems hardly challengeable given its emphasis on concepts such as human rights. However, as such, it is also largely bounded by the terms of the agreements and the funding strings attached to them. Support is indeed available as a priority for fostering the inclusion of groups that are explicitly named in the Sendai Framework for Disaster Risk Reduction and other international agreements to the detriment of groups that are not.

It is therefore important to briefly reflect upon how some groups come to be named and others do not. The Sendai Framework for Disaster Risk Reduction provides an interesting example here. In the months leading up to the signing of the agreement in March 2015, intense lobbying from the representatives of a number of groups disproportionally affected by disas-ters occurred. These representatives included the United Nations Major Groups and Stakeholders but also NGOs which advocate for the rights of specific groups, such as children, people with disabilities, or older people. These advocacy organisations, in very good faith, did the most they could for their groups to be named in the document, so that they could eventu-ally stir attention from national government signatories of the framework and donor organisations that will support the implementation of the agree-ment. This is a fair and laudable process.

However, this lobbying happened in silos with limited overarching coordination and assessment of the broader picture with regard to who was able to speak on behalf of and represent whose groups and the position of the latter in society. As a result, the six groups (women, children, people with disabilities, older people, indigenous people and migrants) that made it into the Sendai Framework for Disaster Risk Reduction reflect how powerful their representation was at the negotiation table. This process inherently leaves behind a larger number of other groups, including people in prison, homeless, and people who do not identify within the female–male gender binary, to just name a few, although these groups suffer disproportionally when dealing with natural hazards. This approach is at odds with one of the key principles of inclusion, which is to leave no one behind.

Similar observations are made on the ground when it comes to actually fostering inclusion. Inclusion often is practiced in silos in that it usually is encouraged for one particular group in isolation from other dominant groups in society: those who are in power in the first place. It is therefore a mechanical, rather than political, process of creating boxes as close as possible to a putative centre. A process that sticks to a view of the world where those groups are not seen as part of a whole (i.e. society) but in isolation. Inclusion therefore is often exclusive.

In practice, it is frequently observed that so-called 'inclusive' projects focus only on women or only on people with disabilities, for example, with very limited interaction from men or people without disabilities, respectively (Figure 7.1). The main exception to the rule is initiatives geared at fostering the inclusion of children, which usually involve parents and teachers (Petal et al., 2020; Wisner, 2006). These initiatives are laudable and have the potential to build upon the organic capacities of these groups. However, they fail to address the unequal power relations that underpin the so-called 'marginal position' of these groups in society, which are at the core of the inclusion agenda in the first place (Chambers, 1983; Maskrey, 1984). In fact, one may argue that members of the 'vulnerable' groups are aware of their capacities and vulnerabilities. The challenge is rather to get these recognised by those with more power so that the latter agrees to relinquish power and grants access to means of protection and resources to those who are to be included. This approach is about recognition rather than identity (Fraser, 1995, 2000).

There are several reasons for the practice of exclusive inclusion. One is that the dominant upward accountability mechanisms instilled by the inclusion imperative in policy, and the accompanying funding strings, favour short-term outcomes over the longer-term transformation of social structures and the redistribution of power and resources within society. We will discuss this issue later in this chapter. In addition, upward accountability mechanisms encourage tangible and measurable outputs and outcomes (see also Chapter 4) to the detriment of the more fluid, intangible and loose nature of transferring power. Finally, the same accountability mechanisms often discourage organisations that facilitate inclusion to seek

ISDR
International Strategy for Disaster Reduction

Gender Perspectives: Integrating
Disaster Risk Reduction into
Climate Change Adaptation
Good Practices and Lessons Learned 2008

Figure 7.1 Cover of a report of the United Nations Strategy for Disaster Reduction
on fostering the inclusion of women in disaster risk reduction

conflicts. Yet the occurrence of conflicts often reflects a genuine process of power-sharing and hence a truly inclusive process.

As a result, the practice of inclusion in disaster risk reduction has widely steered away from its original agenda and ethos. It has become a-political, standardised and normative. As such, it raises a number of issues, the first of which is the delineation of groups and categories.

On categories, intersectionality and power relations

The practice of exclusive inclusion builds upon clear distinctions drawn between groups, especially amongst those assumed to be outside or at the margin of society. The practice of exclusive inclusion, thus relies upon categories such as women, children, people with disability, and older people.

Categories are here '*une des manières pour l'"être" de se dire ou de se signifier*' (Derrida, 1972, p. 218). These categories are both multiple and fluid but they all reflect a taxonomic world view inherited from the Enlightenment. Spivak (1974, p. xxii) writes that

> because he or she [i.e. the human being] must be secure in the knowledge of, and therefore power over, the "world" (inside or outside), the nerve stimuli are explained and described through the categories of figuration that masquerade as the categories of "truth".

Categorisation thus aims at rationalising objects and subjects, place and time, phenomena and processes, and so on. As such, categories lead to reifying and fragmenting the world yet within a reductionist and structuralist perspective in a Lévi-Strauss (1962) sense. Indeed, frequently, these ontological categories ultimately rely on a binary understanding of the world – such as women versus men, children versus adults, people with disabilities versus people without disability, indigenous versus non-indigenous, migrants versus locals, incarcerated versus free individuals, and the homeless versus the housed. Exceptions include contemporary differentiations based on race and ethnicity. In most cases, however, these assumptions reflect Western world views as in the case of the female–male binary as the dominant framing of gender that we explore in Chapter 8.

In dealing with disasters, these groups are usually distinguished on the basis of their vulnerability or capacities/resilience, as per the imperative dictated by the concepts used to understand disasters. These distinctions based on people's abilities to deal with hazards and disasters underpin a process of labelling. If children, women, people with disabilities, and so on are named in the Sendai Framework for Disaster Risk Reduction, it is because evidence, generated through the assumption that such clear categories exist, shows that they are disproportionally affected by disasters. Our intention here is not to challenge the material existence of this fact. Rather, we question the labelling of these groups assumed to exist in isolation, the implication of that labelling, and the power relations that the process of labelling mirrors and entails.

The first point – that is, the ontological existence of distinct categories in isolation from each other – raises the issue of intersectionality. Butler (1995, p. 50) writes that 'identity categories are never merely descriptive, but always normative, and as such, exclusionary'. Categories and labels therefore put people in narrow and exclusive boxes. These boxes do not allow for the possibility that a woman can live with disabilities and be of older age. As Crenshaw (1989, pp. 166–167) wrote:

> the failure to embrace the complexities of compoundedness is not simply a matter of political will, but is also due to the influence of a way of thinking about discrimination which structures politics so that struggles are categorized as singular issues.

The inherent nature of intersectionality has received widespread attention in many fields of study (e.g. Carbado et al., 2013; Cho et al., 2013). It initially developed in the 1980s around the intersection between race, class and gender (Crenshaw, 1989, 1991; Spivak, 1987, 1988) but has since expanded to many forms of possible intersections across age, disability, housing, and so on, including in disaster studies, and through the practice of disaster risk reduction (e.g. Chaplin et al., 2019; Schuller, 2015; Vickery, 2018).

Further concerns emerge when the labels assigned to specific groups turn into rigid stereotypes, such as 'Children and older people are all vulnerable to disasters' or, in Butler's words (1995, p. 50), when 'the signified has been conflated with the referent'. This concern is particularly salient when the labels are disempowering and stigmatising, as in the case of the nexus vulnerability–marginality. Then, labelling becomes 'an act of valuation and judgment involving prejudices and stereotyping' (Wood, 1985, p. 348). As Butler (2016, pp. 24–25) adds:

> once groups are marked as "vulnerable" within human rights discourse or legal regimes, those groups become reified as definitionally "vulnerable," fixed in a political position of powerlessness and lack of agency. All the power belongs to the state and international institutions that are now supposed to offer them protection and advocacy. Such moves tend to underestimate, or actively efface, modes of political agency and resistance that emerge within so-called vulnerable populations.

As such, the labelling of certain groups as 'vulnerable' justifies the 'intervention' of outside actors who are to uplift them from their predicament. When this discourse crosses cultural boundaries and applies to the non-Western 'other', those who suffer from disasters in the 'dangerous South', it prolongs an imperialist legacy captured in Bhabha's (1994, pp. 118–119) powerful critique:

> racist stereotypical discourse, in its colonial moment, inscribes a form of governmentality that is informed by a productive splitting in its constitution of knowledge and exercise of power. Some of its practices recognize the difference of race, culture and history as elaborated by stereotypical knowledges, racial theories, administrative colonial experience, and on that basis institutionalize a range of political and cultural ideologies that are prejudicial, discriminatory, vestigial, archaic, 'mythical', and, crucially, are recognized as being so. By 'knowing' the native population in these terms, discriminatory and authoritarian forms of political control are considered appropriate.

Many actors of disaster risk reduction thus use labels to 'quantify and measure categories of people to define their needs, justify interventions and to formulate solutions to perceived problems' (Moncrieffe and Eyben, 2007, p. 1). Labelling then becomes a matter of numbers and a marketing strategy where politically loaded concepts such as vulnerability and marginality become buzzwords or 'a taxonomic diagnosis in our trade' (Spivak, 1990c, p. 239). Here, intervention is to be differentiated from that of those who block access to resources and means of protection in dealing with natural hazards. By intervention, we mean the application of the normative and regulatory precepts of governmentality that we discussed in Chapter 5 and that justifies the contribution of a number of actors by the inability of the 'marginalised' to act by themselves. Labelling therefore has both 'classificatory and regulatory functions' (Moncrieffe and Eyben, 2007, p. 1).

Labelling some groups as vulnerable or with more or less capacities or as resilient or not is therefore a matter of power relations. However, the question is not so much about the existence of categories (Butler, 1995) or labels (Moncrieffe and Eyben, 2007; Wood,1985) per se, as these seem inherent to most societies and human relations, but rather 'which labels are created, and whose labels prevail to define a whole situation or policy area, under what conditions and with what effects?' (Wood, 1985, p. 349). The question is particularly salient when some groups are labelled 'vulnerable', especially when vulnerability stands alongside disempowering concepts such as marginality, and others are not, whether they disproportionally suffer from natural hazards or not. It is therefore a matter of who labels whom – and why.

Since vulnerability and other concepts used to understanding disasters are concepts that more often than not are alien to many diverse local cultures around the world, labelling through these concepts is predominantly a re-presentation of people's own conditions and identities. There are therefore inherent asymmetrical power relations between the labellers and the labelled (Moncrieffe and Eyben, 2007). As Wood (1985, p. 347) writes, 'labelling refers to a relationship of power in that the labels of some are more easily imposed on people and situations than those of others. It is therefore an act of politics involving conflict as well as authority'. Since labelling often supports a process of lobbying in policy, as discussed in the previous section, re-presentation is conflated with representation. Who speaks on behalf of whom and who is powerful enough to be represented are therefore crucial. A point that raises two underlying issues: one of legitimacy and one of leverage.

Who is legitimate to speak on behalf of whom in disaster risk reduction is contested. It largely reflects one's positionality and identity, two traits that are fluid, subjective and sensitive. For example, whether Western, white and well-off women, to caricature, are able to speak on behalf of women who have very different concerns across Latin America, Africa, Asia and the Pacific has stimulated heated debates amongst feminists (e.g. Lugones, 2007; Mohanty, 1984; Oyěwùmí, 1997). In many instances, it has led to what Said called possessive exclusivism or

> the sense of being an excluding insider by virtue of experience (only women can write for and about women, and only literature that treats

women or Orientals well is good literature), and second, being an excluding insider by virtue of method (only Marxists, anti-Orientalists, feminists can write about economics, Orientalism, women's literature).

(Said, 1985, pp. 106–107)

Nowadays, positionalities are further blurred by globalising processes, especially migration patterns that challenge place-based identities, so that living in one place or being from another does not necessarily ensure legitimacy, nor does it preclude it. As Chakrabarty (2000, p. 1) once said, 'passports and commitment blur the distinctions of ethnicity in a manner that some would regard as characteristically postmodern'.

The question of leverage is somehow less touchy yet similarly critical. It is a matter of who is powerful enough to stir enough attention to gain recognition on the political scene and hence in policy. Wood (1985, p. 358) actually argues that 'the function of labelling is to achieve universalistic justification or legitimation of something which is highly selective. Labels require advocacy, therefore, and successful advocacy is accompanied by; that is, a proliferation of practices in support of the prevailing orthodoxy'. For example, those groups that are named in the Sendai Framework for Disaster Risk Reduction and their representatives, especially NGOs, are assured of receiving significant commitment and support from governments and donors which are expected to meet the requirements of the agreement. This is obviously a positive step forward should these groups claim such attention and support. As Fraser (2000, p. 109) recognises, 'properly conceived, struggles for recognition can aid the redistribution of power and wealth and can promote interaction and cooperation across gulfs of differences'.

Yet the attention that some groups gain by being named in policy inherently excludes those other groups that have not made it to the same level of recognition. Indeed, if you are not identified on the political scene, then you are likely to become invisible or 'represent an outside beyond the boundaries of political society' (Chatterjee, 2008, p. 61). The reasons for such a lack of leverage on the political and policy scenes are multiple. They include a 'lack of knowledge of their existence and people's own inability to mobilize, gain access to the right networks and to position their issues in sufficiently commanding and persuasive ways' (Moncrieffe and Eyben, 2007, p. 3). Ultimately, the discourse and practice of inclusion are therefore exclusionary and further 'marginalising' and, along the lines of neoliberalism, encourage competition amongst groups rather than solidarity with others. This discourse and practice of inclusion define priorities and rank groups 'through the distribution of status and worth' and dictate 'the rules of access to particular resources and privileges' (Wood, 1985, p. 349, 352).

These foregoing concerns should not undermine the fact that categories, especially, and labels, by extension, are constitutive of human interactions and support political struggles (Wood, 1985; Butler, 1995). Crenshaw (1991, pp. 1296–1297) writes:

> to say that a category such as race or gender is socially constructed is not to say that that category has no significance in our world. On the contrary, a large and continuing project for subordinated people – and indeed, one of the projects for which postmodern theories have been very helpful – is thinking about the way power has clustered around certain categories and is exercised against others. This project attempts to unveil the processes of subordination and the various ways those processes are experienced by people who are subordinated and people who are privileged by them. It is, then, a project that presumes that categories have meaning and consequences. And this project's most pressing problem, in many if not most cases, is not the existence of the categories, but rather the particular values attached to them and the way those values foster and create social hierarchies.

As a result, our point is not to suggest that categories should no longer be used in disaster studies or in disaster risk reduction. Rather, our point is 'to relieve the category of its foundationalist weight in order to render it as a site of permanent political contest' (Butler, 1995, p. 41). Therefore, categories and labels should be constantly challenged and reconsidered, especially when they are cast through a Western lens (Young, 2004). In fact, they 'should be defined and grounded in the local milieu, rather than based on "universal" findings made in the west' (Oyěwùmí, 1997, p. 16) and ultimately we should recognise that they are just 'invention of our own necessity' (O'Hanlon and Washbrook, 1992, p. 145).

Myths and realities: of communities and local knowledge

The practice of exclusive inclusion is further perpetuated through two myths that have served formative purposes: one is the myth of 'community' and the other is the romanticising of local/indigenous/traditional knowledge. These two myths further reflect how the discourse on inclusion in disaster risk reduction is grounded in Western ontologies and epistemologies.

On the myth of community

In practice, in line with Maskrey's (1984) original agenda, fostering inclusion in disaster risk reduction through participation is often enacted as part of so-called 'community-based/led/managed' initiatives which imply a universal concept of community. The myth that there are such things as universal communities has been widely criticised and revisited, including in disaster risk reduction (Cannon, 2014; Titz et al., 2018). This critique revolves around the assumption that community entails 'a homogeneous group of people sharing mutual interests, or a social network merely defined by location' (Titz et al., 2018, p. 3). In fact, Maskrey (1984, p. 28), in his agenda-setting paper almost 40 years ago, flagged it as a trap not to fall into:

> The distinction between "community" and "locality" is all too frequently blurred. It is wrong to presuppose that the physical grouping of people together in a given locality automatically constitutes a community. Although community would seem to imply a geographically limited area, its essential characteristic is always the existence of some form of social organization.

Indeed, community is another of these Western concepts that have pervaded the lingua of disaster studies and disaster risk reduction. It stems from the Latin *communitas* and is believed to have travelled into English around the fourteenth century. Today's use of the concept in disaster risk reduction is, in fact, very similar to that captured in Diderot and d'Alembert's *Encyclopédie* back in the eighteenth century:

> le corps des habitans d'une ville, bourg, ou simple paroisse, considérés collectivement pour leurs intérêts communs. (…) l'objet de cette communauté consiste seulement à pouvoir s'assembler pour délibérer de leurs affaires communes, & avoir un lieu destiné à cet effet[3].

Its contemporary meaning, however, owes much to Tönnies's foundational binary distinction between *Gesellschaft* (roughly translated as society) and *Gemeinschaft* (community) that he made from his studies of Germany at the close of the *Aufklarung*. In Tönnies's (1887, p. 16) own words:

> Denn die Gemeinschaft des Blutes, als Einheit des Wesens, entwickelt und besondert sich zur Gemeinschaft des Ortes, als welche im Zusammen-Wohnen ihren Ausdruck hat, und diese wiederum zur Gemeinschaft des Geistes als dem blossen Miteinander-Wirken und Walten in der gleichen Richtung, im gleichen Sinne. Gemeinschaft des Ortes kann als Zusammenhang des animalischen, wie die des Geistes als Zusammenhang des mentalen Lebens begriffen werden, die letztere daher, in ihrer Verbindung mit den früheren, als die eigentlich menschliche und höchste Art der Gemeinschaft[4].

Thus, *Gemeinschaft* referred to a small aggregate of people, often located away from centres of power but having a common and continuous way of life, similar beliefs, close ties, trust and frequent interactions. As such, the concept reflects not only the reality of a particular place and time (Germany in the early nineteenth century) but also the epistemology associated with this particular context, which was emerging from the Enlightenment. This understanding of community is appealing to an approach of disaster risk reduction which emphasises local people and their capacities/

resilience for several reasons. First, it refers to local realities and local people at the centre of the inclusion agenda. Second, the collective nature of the concept appears conducive for promoting an approach to disaster risk reduction where people are meant to collaborate and share resources, knowledge and skills. As such, it is cost-effective and strengthens the first line of defence in dealing with disasters. It also reinforces the responsibility of individuals to the detriment of the state. As a result, it constitutes a powerful narrative in response to the critique that disaster risk reduction has long been a top-down and technocratic affair.

The lingering use of the concept of community in this perspective directly inherited from the Western academic tradition has been widely criticised for failing to capture local realities in diverse regions of the globe (Agrawal and Gibson, 1999; Blaikie, 2006; Cleaver, 1999; Gujit and Shah, 1998; Moser, 1989). In his famous foreword to Guijt and Shah's book entitled the *Myth of Community*, Chambers (1998, p. xviii) writes, in the specific context of gender:

> local contexts are complex, diverse and dynamic. The reductionism of collective nouns misleads: 'community' hides many divisions and differences, with gender often hugely significant; 'women' as a focus distracts attention from gender relations between women and men, and from men themselves; and 'women' also conceals the many differences between females by age, class, marital status and social group. Nor are common beliefs valid everywhere: female-headed households are often the worst off, but not always. Moreover, social relations change, sometimes fast. It is not just the myth of community that this book dispels, but other myths of simple, stable and uniform social realities.

The myth of community thus denies both the existence of different interests and unequal power relations at the very local level as well as the inherent interactions of the local with the global. The former may be explicit, as in caste or peasant societies where resources are unequally distributed on the basis of who you are or whether you own land or not, or more subtle and embedded within the normative and regulatory approach to governmentality discussed in Chapter 5. The latter involves flows of capital and commodities as well as less tangible fluxes associated with diverse forms of remittances and interpersonal communications. The interactions between the local and the global also blur the distinction that the concept of community entails between an inside and an outside.

These well-known limitations of the concept of community, as embedded in the dominant discourse on community-based/led/managed disaster risk reduction, have two main implications for the practice of inclusion. First, assuming homogeneity often leads to overlooking intrinsic inequalities and unequal power relations that, in theory, are the bottom line of the process of inclusion. Rather, it often ends up favouring those already in power. Williams (2004, p. 562) powerfully writes that by homogenising differences within communities, and uncritically privileging 'the local' as the site for action, far too many accounts of participatory development are in danger of actively de-politicising development. They draw a veil over repressive structures (of gender, class, caste and ethnicity) that operate at the micro-scale but are reproduced beyond it and, by emphasising 'the community' as the site where authentic development can occur, they direct attention away from wider power relationships that frame local development problems.

Moreover, the assumption that there are well-defined and homogenous communities mirrors the presumed existence of clearly distinct categories, as discussed in the previous section of this chapter. As such, community-based/led/managed disaster risk reduction and its inclusive agenda, despite Maskrey's (1984) original warning, often fail to address the intersecting and multidirectional relationships between the alleged 'outside'/ global and the so-called 'inside'/local. It leads to decontextualising and isolating the local, as suggested by Freire (1968, IV, pp. 24–25):

> nos trabalhos de "desenvolvimento de comunidade", sem que estas comunidades sejam estudadas como totalidades em si, que são parcialidades de outra totalidade (área, região, etc.) que, por sua vez, é parcialidade de uma totalidade maior (o país, como parcialidade da totalidade continental) tanto mais se intensifica a alienação. E, quanto mais alienados, mais fácil dividi-los e mantê-los divididos. Estas formas focalistas de ação, intensificando o modo focalista de existência das massas oprimidas, sobretudo rurais, dificultam sua percepção critica da realidade e as mantém ilhadas da problemática dos homens oprimidos de outras áreas (...).[5]

One obvious example of such interactions that matter is the rapport between migrants and their homeland or, conversely, those of tourists with their place of residence. One cannot omit such relationships, in either direction, when devising disaster risk reduction. Migrants provide support to their homeland in the same way as the homeland worries about the migrants when they are affected. Similarly, tourists are affected by local disasters as much as their relatives worry about them, and vice versa. Should there be such a thing as communities, they should then be considered beyond propinquity (Webber, 1964).

The core assumption, inherited from a Western concept identified in the eighteenth and nineteenth centuries, that there are such things as communities that are bound to a small and well-defined place, have a common and homogenous social structure, and have common interests and shared norms (Agrawal and Gibson, 1999) thus not only is outdated but also fails to depict the reality of settlements, social structures, values and norms outside of Europe. Its underpinning role in sustaining a romanticised neocolonial agenda focused on the local, the rural and the people is well known (Chambers, 1974; Mayo, 1975). So are its biases and shortcomings (Cleaver, 1999; Cooke and Kothari, 2001; Williams, 2004).

On romance and local/indigenous/traditional knowledge

The discourse and practice of inclusion in disaster risk reduction further build on the observation that people are not helpless in dealing with natural hazards and that they hold very useful forms of knowledge. This knowledge is grounded, useful in preventing disasters and embedded within people's everyday lives. Therefore, pulling such knowledge within disaster risk reduction, at least its formal declension as bounded by Western standards, is meant to bring relevance, enhance ownership and, as a result, foster meaningful participation and the inclusion of local people in disaster risk reduction.

It is not our intention to challenge that people are not helpless and that they hold very useful forms of knowledge. In fact, it underpins our argument that the people who hold this knowledge are the ones who should decide what a disaster is in their eyes and what should be done to reduce risk if there is risk.

Rather, our point is to follow Agrawal (1995) and contest that there is a romanticised divide that polarises what is known as scientific/Western knowledge and the knowledge of people at risk, whether that is labelled local, traditional, indigenous or whatever else. Notably, Agrawal does not dispute the existence of different forms of knowledge per se. Rather, he challenges the fabricated divide between scientific/Western knowledge and other forms of knowledge known as local/indigenous/traditional and the binary view of the world that such a divide casts. This binary view of the world mirrors asymmetrical power relations, which have allowed for one tradition of knowledge to dominate over a myriad of others and for the latter to become subjected to the norms of the former. Foucault (1997, pp. 8–9) further explains that

> par "savoirs assujettis", j'entends également toute une série de savoirs qui se trouvaient disqualifiés comme savoirs non conceptuels, comme savoirs insuffisamment élaborés: savoirs naïfs, savoirs hiérarchiquement inférieurs, savoirs en dessous du niveau de la connaissance ou de la scientificité requises[6].

This alleged hierarchy of knowledges has sustained interventions of the West in the rest of the world in the name of development (Agrawal, 1995; Bhabha, 1994; Escobar, 1995).

Our argument is that there are indeed multiple forms of knowledge but that they should not be seen in a dialectical perspective with scientific knowledge. Rather, the latter should be seen as one, yet hegemonic, form of knowledge, among many others. Furthermore, the divide between the many others and the hegemonic one is not as clearly marked as argued by the advocates of local/traditional/indigenous knowledge. Here, we shall expand our point using Agrawal's (1995) triple line of argumentation, which is that the divide between scientific/Western and local/indigenous/traditional knowledges is fabricated on the basis of substance, epistemology and contextuality.

First, it is usually assumed that scientific knowledge is concerned with abstract realities and the drawing of grand analytical theories while local/indigenous/traditional knowledge pictures everyday life and people's livelihoods. We agree with Agrawal that, both in general terms and in dealing with disasters, such a divide is irrelevant. Multiple studies show that local/indigenous/traditional knowledge is also preoccupied with understanding the world at large and how disasters fit within it (e.g. Balay-as, 2019; Thufail, 2019) whereas scientific knowledge very much applies to solving issues that affect people's daily lives, including the very occurrence of natural hazards and disasters.

Second, it is obvious and core to our argument that any different forms of knowledge, whether labelled scientific/Western or local/indigenous/traditional, mirror diverse ontologies and epistemologies. However, it is untenable to consider that scientific/Western knowledge stands aside from all other forms of knowledge for its unique rational, systematic, analytical and iterative nature. The very famous case of the people of Simeulue shows that they had built upon the experience of the 1907 tsunami to expand and enhance their knowledge of a particular natural phenomenon in order to analyse the environmental signals that forewarned the tsunami that struck in 2004. The people of Simeulue then took the rational decision to evacuate to the nearby hills in a very organised fashion (McAdoo et al., 2006). On the other hand, the rationality and universality of Western science have long been challenged by the likes of Feyerabend (1975, 1987) and Latour and Woolgar (1979).

Third, scientific/Western knowledge is assumed to be of universal relevance, in both time and space, whereas local/indigenous/traditional is relevant only in a specific location at a given time. We have seen in the previous paragraph that local/indigenous/traditional is never stuck in time. It constantly evolves in the same way as scientific/Western knowledge. Moreover, scientific/Western knowledge is not of universal relevance, or there would be no reason to write this book.

The foregoing three points challenge the ontological existence of a knowledge binary. That binary would oppose two clearly distinct forms of knowledge that are based on very different substance, epistemologies and contextuality. In fact, it is not only the binary that is to be contested but also the apparently strict and impermeable partition between the two sides of the divide. Indeed, literature abounds in examples of either hybrid forms of knowledge or places where different forms of knowledge co-exist and of people who combine or selectively choose from these forms of knowledge to make informed decisions either to deal with immediate and practical issues or to make sense of metaphysical concerns (e.g. Balay-as, 2019; King and Goff, 2010; Thufail, 2019).

Ultimately, the alleged divide between scientific and local/indigenous/traditional knowledge is fabricated to fit the ontological divide between nature/hazard and culture/vulnerability. In the practice of disaster risk reduction, scientific knowledge is predominantly positioned towards the hazard while local/indigenous/traditional knowledge is assumed to reflect the cultural side of the binary (including its interpretation of the hazard). As such, the alleged divide between knowledges expands and reinforces the ontological assumption that disasters sit at the interface between nature and culture. Furthermore, one may argue that the increasing attention that local/indigenous/traditional knowledge has received over the past couple of decades mirrors the Hegelian dialogue discussed in multiple instances in this book, whether it was with regard to scientific paradigms, methods for assessing disaster risk, or strategies to reduce it. The existence of such dialogue supports our argument that Western ontologies and epistemologies pervade all dimensions of the *dispositif* deployed in dealing with what we call disaster.

In fact, it is easy to extend the argument to the most recent push for integrating these two alleged forms of knowledge as the ultimate step in Hegel's quest for reason. This step would be inclusive and cognisant of everyone's perspective – a sort of forced hybridisation of knowledge that, we argue here, is doomed to fail. For it seems epistemologically impossible and incoherent to pull together different world views/senses that have nothing in common, including the binary view of the world that underpins the Western discourses on disasters. In practice, such a process of integration, when it leads somewhere, is to the absorption and dilution of the expression of the so-called local/indigenous/traditional forms of knowledge, in all their diversity, within the norms and rules of scientific knowledge. Agrawal (2002) calls this process *scientisation*, wherein bits and pieces of local/indigenous/traditional knowledge useful to disaster risk reduction are singled out of their context (*particularisation*), examined and tested against scientific criteria (*validation*) and then shared more broadly (*generalisation*). Agrawal (2002, p. 291) adds that 'statements that are successfully particularised, validated, and generalised become knowledge by satisfying a particular relationship between utility, truth, and power'. Through this process, local/indigenous/traditional knowledge is subjected to the norms of scientific knowledge (Foucault, 1997) and becomes the alibi for a token inclusive approach of disaster risk reduction.

Process, means and the dilemma of accountability

Ultimately, the practice of inclusion in disaster risk reduction is embedded within the overarching modern art of government that is governmentality. We discussed the overall principles and *dispositif* that support the latter, including in integrating the discourse around community-based/led/managed disaster risk reduction, in detail in Chapter 5. Here, we turn to its implication for the practice of inclusion and how its normative and regulatory nature has led to skew the original ethos of the concept and discourse, as articulated in the 1980s and outlined earlier in this chapter.

The fixation on categories and labels, in particular, has been central to the normalisation of the discourse on inclusion and the standardisation of its application in practice. Labels applied to 'vulnerable' groups have become boxes to tick and requirements in project proposals and reports outside of their cultural and political context. These labels have become the basis for reductionist risk assessment matrixes and frameworks as identified in Hazard, Vulnerability and Capacity (HVCA) toolkits, which we discussed in Chapter 4. These templates and frameworks are to be applied all around the world with only minor tweaking allowed to supposedly recognise local culture.

Individuals have also become numbers that justify inclusion and participation through attendance sheets and photos showing crowds of sorts. The 'number game' has conflated involvement and participation or involvement and inclusion, where the number of 'participants' and their diverse labels have become more important than the process of participation and sharing of power. In this perspective, participation and inclusion have become means rather than processes. The importance of the means over the process has been driven by the bureaucratisation and de-politicisation of inclusion and participation as well as the demand of donors and government agencies that expect standards and indicators identified in project proposals to be met (Cornwall and Brock, 2005). These standards and indicators are designed after the requirements of international agreements discussed in Chapter 5. As Edwards and Hulme (1995, p. 12) were anticipating, 'given the financial and political muscle of official agencies, there is an obvious fear that donor funding may reorient accountability upwards, away from the grassroots, and bias performance-measurement toward criteria defined by donors'. Twenty-five years later, the inherent downward nature of accountability in participatory and inclusive initiatives has indeed vanished to the benefit of upward accountability.

Both the regulatory nature of the *dispositif* and the drive for upward accountability have had two major implications in practice. First, these two tenets of governmentality have favoured an approach that is bound by the terms of international agreements, such as those of the Sendai Framework for Disaster Risk Reduction. This approach encourages a focus on plans, outlined in time and space, as discussed in Chapter 5, and their design through 'projects' or initiatives, also bound in time and space, and designed to meet specific objectives aligned with the expectations of international agreements and those of donors and government agencies. In many contexts, these objectives and expectations unfortunately differ from local realities, as we saw in the previous chapter in the case of Kiribati.

Furthermore, it is the expectation of the regulatory nature of the *dispositif* and its concomitant demand for upward accountability that the objectives of these projects are met. NGOs and local government agencies that implement such projects are expected to deliver what has been promised in proposal or in policy, respectively. Then, the complex, structural and conflictual dimensions of power relations that underpin the discourse and practice of inclusion and that require the protracted transformation of power structures within society become obstacles rather than pathways. As a result, local practitioners frequently avoid, rather than embrace, them (Chambers, 1998; Cleaver, 1999). Therefore, the sharing of power through participation that is expected, in theory, to achieve genuine and sustainable inclusion is bypassed to the benefit of decisions that ensure that the project objectives are met. The process of participation and inclusion is therefore *facipulated*, and outcomes take priority over the process of participation.

This approach to inclusion geared towards risk assessments, plans, objectives, deliverables, outcomes and projects epitomises the ethos and principles of the project of modernity of the West, which is to both control nature and rationalise society to allow people to live a life free of dangers. If, in theory, we acknowledge that the discourse on inclusion offers room for a radically different perspective by offering local people the opportunity to identify both what threatens their lives and what should be done to deal with these hazards, the actual practice of inclusion does not significantly depart from the seemingly different technocratic and military-inspired approach to disaster risk reduction. If we follow Foucault (2004a), we may actually argue that the shift from one side of the binary to the other, from a focus on nature and hazards to another on culture and people, mirrors a transition from a *dispositif* of explicit discipline to one of security where disciplines and regulations are more insidiously embedded within the art of government.

It is therefore safe to argue that the potentially radical shift in our approach to disaster risk reduction entailed by the emergence and mainstreaming of the discourse on, and practice of, inclusion has, in reality, perpetuated the hegemony of Western understandings of disaster. The packaging has changed but the transition on the ground has been from a normative top-down approach to one that is normative from the 'bottom up' or allegedly so. Thus, it is easy to agree with Bhabha (1994, p. 347) that 'there is the danger that the mimetic contents of a discourse will conceal the fact that the hegemonic structures of power are maintained in a position of authority through a shift in vocabulary in the position of authority'. This continuity in the way we approach disaster risk reduction has been sustained by the continuing hegemony of Western scientific paradigms in framing the discourses on disasters, whether these are centred on hazard or vulnerability. Indeed, fully embracing the ethos of inclusion in disaster risk reduction inherently requires us to step out of the nature–culture binary or rather to consider this binary solely in contexts where it makes sense from an ontological and epistemological perspective, which is not in all regions of the world.

There are, of course, exceptions to the rule (and fortunately so) and examples that the transformation of power relations through genuine participation is possible and that it can indeed lead to inclusion and empowerment, both in the West and in other regions of the world in their diversity (e.g. Bhatt, 2014; Delica, 1999; Maskrey, 1989; Norton and Gibson, 2019; Tsunozaki, 2012). Many of these examples of genuine inclusion are organic and initiated by those who claimed inclusion in the first place rather than outsiders (e.g. Bajek et al., 2008; Bhatt, 2017; Carlton and Mills, 2017; Kelman and Karnes, 2007; Kenney and Phibbs, 2015). The Philippines provides an interesting example of widespread, though far from systematic, initiatives that have fostered genuine inclusion in disaster risk reduction embedded within a broader agenda to transform power relations within society (e.g. Abinales, 2012; Delica, 1999; Heijmans and Victoria, 2001; Luna, 2003; Orejas, 2003). Spearheaded by a tight network of civil society organisations formed out of the struggle against the dictatorship of Ferdinand Marcos in the 1970s and 1980s, disaster risk reduction has long been seen as an opportunity and pathway towards addressing skewed power relations that cause what locals have been calling disasters. These include a broad range of issues associated with natural hazards but also with lingering food insecurity, the depletion of natural resources, and unequal access to land and other resources (Delica, 1993; Heijmans, 2012; Luna, 2001).

Is inclusion culturally ethical?

Suggesting that the practice of inclusion is most often skewed towards or by the ethos and principles of the project of modernity is one thing. Another is to take a further step backward and actually question the very nature, origin and relevance of the discourse on inclusion as pitched in disaster risk reduction and other dimensions of society. This questioning is about whether it is actually ethical, as in moral from the Greek τά ἠθικά, and coherent from a Western perspective to foster the inclusion of those who are believed to live at the margins of society and who are disproportionally affected by what the same West calls disaster. Indeed, in many instances, encouraging the participation of those the West considers at the margins of society requires us to challenge and transform existing cultural norms and values. Yet the same international agreements and national policies, such as the Sendai Framework for Disaster Risk Reduction, that have mainstreamed the discourse on inclusion also emphasise that culture – and, more specifically, traditions – should be of paramount importance in disaster risk reduction (Mercer et al., 2010; Shaw et al., 2009). In fact, the drive to preserve and pull local/indigenous/traditional knowledge into disaster risk reduction is contradictory if we agree that this form of knowledge is fully embedded within cultures, which may, in Western eyes, be discriminatory to some individuals.

For example, fostering the inclusion of women in disaster risk reduction in India or Nepal requires dealing with the caste system. Similarly, in Australia, fostering the inclusion of women among Aboriginal groups may require a challenge to the knowledge and power of traditional owners of the land, who, at the same time, are increasingly recognised as the key stakeholders of local development. In both cases, local women may spontaneously advocate for their own rights and desires, which reflects a genuine, organic and grounded momentum. However, when such momentum is induced by outsiders, especially Western stakeholders, this may become a challenge. Why should Western values and norms prevail over local culture?

One may answer that the inclusion of those said to be at the margins of society builds upon evidence that these groups are disproportionally affected in disasters and therefore that they deserve particular attention. Of course, we do not deny the material reality of this fact but we also need to consider (as discussed in Chapters 1 and 4) that both our understanding of what a disaster is and how we provide evidence of their occurrence and impact are skewed by Western epistemologies.

The other obvious argument is that fostering the inclusion of those groups in disaster risk reduction is a matter of human rights, as per the Universal Declaration of Human Rights. These human rights, in the field of disaster risk reduction, are taken on and rolled out by international stakeholders, whether international organisations or NGOs, in the name of their independence and neutrality. This is a fair call but these universal human rights are directly inspired by the French *Déclaration des Droits de l'Homme et du Citoyen* of 1789 and therefore reflect a Western tradition. These universal human rights have become a *métarécit*, which is firmly grounded in the heritage of the Enlightenment (Levinas, 1987). This *métarécit* has been challenged for being a tool, allowing the West to advance its imperialist agenda around the world (de Sousa Santos, 2002; Donnelly, 2013). As Mbembe (2013, p. 25) powerful summarised:

> Dans son avide besoin de mythes destinés à fonder sa puissance, l'Hémisphère occidental se considérait comme le centre du globe, le pays natal de la raison, de la vie universelle et de la vérité de l'humanité. Quartier le plus "civilisé" du monde, l'Occident seul avait inventé un "droit des gens". Seul, il était parvenu à constituer une société civile des nations comprise comme un espace public de réciprocité du droit. Seul, il était à l'origine d'une idée de l'être humain possesseur de droits civils et politiques lui permettant de développer ses pouvoirs privés et publics comme personne, comme citoyen appartenant au genre humain et, en tant que tel, concerné par tout ce qui est humain[7].

In fact, calls for the Declaration of Human Rights to consider local culture, values and norms have remained unanswered and the *dispositif*, still in Foucault's terms, that promotes human rights has been remarkably durable (Donnelly, 2013; Goodale, 2006; Ignatieff, 2003).

Let's be clear: it is obviously not our intention to argue against the fundamentals of the Declaration of Human Rights. Rather, what is of concern is the proclamation as universal of a particular approach to human rights that is inherited from a particular place and time, namely Europe in the late eighteenth century. This imperialist approach to human rights indeed has excluded the consideration of other perspectives from elsewhere in the world. If we follow Lyotard (1993) and consider that the ability and entitlement to speak are the most fundamental human rights and under-pin inclusion as per our foregoing argument, then listening to other and others' perspectives is fundamental. So that imposing a particular dis-course on human rights on disaster risk reduction or on inclusion upon someone else becomes a violation of such rights. As Lyotard (1993, p. 146) adds, 'the discontent from which contemporary societies are suffering, the postmodern affliction, is this foreclosure of the other'.

In fact, there is evidence of culturally grounded, and probably more pro-gressive in Western terms, pronouncements of human rights that predate the French *Déclaration des Droits de l'Homme et du Citoyen* of 1789. For example, the *Manden Kalikan* or *Manden Charter*, also known as *Serment des Chasseurs*, was allegedly proclaimed sometime in the early thirteenth century (hence roughly at the same time as the *Magna Carta Libertatum* of King John of Eng-land) as a foundational act for the then Mali Empire (Tata Cissé and Sag-ot-Duvauroux, 2003). In all its different iterations, first translated into French, this charter explicitly refers to hunger (and, in some versions, to drought):

Article 5:
Donsolu ko:
Ko gòngò ma niyi
Ko dyònnya ma nyi;
Ko gòngò ni dyònnya nyòkòn ko dyugu tè,
Dunya-so yan.
Ko ka ton ni kala to annu bolo,
Ko gòngò tè mòkò faka tukun, Manden,
Ni dyaa kera na-fèn di.[8]

(version reported by Tata Cissé, 2003)

Human rights are a sensitive and contested issue and, again, it is not our intention to argue that those groups that suffer more in dealing with what we call disasters should not be included in disaster risk reduction. We argue for their inclusion. The point is that the foregoing challenges and con-ditions need to be carefully thought through and grounded in their cul-tural context when advancing a genuine inclusion agenda that recognises underlying unequal power relations in society. This recognition can hap-pen only when those groups that suffer more take the lead of a politically loaded agenda. This agenda is not a mechanical process of integration or recentring. As Freire (1968, III, p. 30) famously wrote, '*não posso pensar pelos outros nem para os outros, nem sem os outros. A investigação do pensar do povo não pode ser feita sem o povo, mas com ele, como sujeito de seu pensar*'[9].

Notes

1 '*because liberating action is dialogical in nature, dialogue cannot be a posteriori to that action, but must be concomitant with it. And since liberation must be a perma-nent condition, dialogue becomes a continuing aspect of liberating action*'. From the

English edition translated by Myra Bergman Ramos and published by Penguin Education in 1996.

2 '*the truth is, however, that the oppressed are not "marginal", are not people living "outside" society. They have always been "inside" – inside the structure which made them "beings for others". The solution is not to "integrate" them in to the structure of oppression, but to transform that structure so that they become "beings for themselves"'*. From the English edition translated by Myra Bergman Ramos and published by Penguin Education in 1996.

3 '*the group of inhabitants of a town, village or mere parish, considered collectively for their common interests. (…) The aim of this community consists in being able to come together to deliberate upon their common issues, and to have a place for this purpose*'. Our definition.

4 '*The Gemeinschaft by blood, denoting unity of being, is developed and differentiated into Gemeinschaft of locality, which is based on a common habitat. A further differentiation leads to the Gemeinschaft of mind, which implies only co-operation and co-ordinated action for a common goal. Gemeinschaft of locality may be conceived as a community of physical life, just as Gemeinschaft of mind expresses the community of neutral life. In conjunction with the others, this last type of Gemeinschaft represents the truly human and supreme form of community*'. From the English edition translated by Charles P. Loomis and published by Harper Torchbooks in 1957.

5 '*In "community development" projects the more a region or area is broken down into "local communities", without the study of these communities both as totalities in themselves and as parts of another totality (the area, region, and so forth) – which in its turn is part of a still larger totality (the nation, as part of the continental totality) – the more alienation is intensified. And the more alienated people are, the easier it is to divide them and keep them divided. These focalized forms of action, by intensifying the focalized way of life of the oppressed (especially in rural areas), hamper the oppressed from perceiving reality critically and keep them isolated from the problems of oppressed women and men in other areas*'. From the English edition translated by Myra Bergman Ramos and published by Penguin Education in 1996.

6 '*When I say "subjugated knowledges" I am also referring to a whole series of knowledge that have been disqualified as nonconceptual knowledges, as insufficiently elaborated knowledges: naive knowledges, hierarchically inferior knowledges, knowledges that are below the required level of erudition or scientificity*'. From the English edition translated by David Macey and published by Picador in 2003.

7 '*In its avid need for myths through which to justify its power, the Western world considered itself the center of the earth and the birthplace of reason, universal life, and the truth of humanity. The most "civilized" region of the world, the West alone had invented the "rights of the people." It alone had succeeded in constituting a civil society of nations understood as a public space of legal reciprocity. It alone was at the origin of the idea that to be human was to possess civil and political rights that allowed individuals to develop private and public powers as citizens of the human race who, as such, were shaped by all that was human*'. From the English edition translated by Laurent Dubois and published by Duke University Press in 2017.

8 *Article 5:*
The children of Sanènè and Kòntròn state that:
hunger is not a good thing,
servitude is not a good thing either;
there is nothing worse than those things in this poor world.
As long as we will hold the quiver and the bow,
hunger will not kill anyone in the Manden.
(Our translation from the French version by Tata Cissé, 2003)

9 '*I cannot think for others or without others, nor can others think for me*'. From the English edition translated by Myra Bergman Ramos and published by Penguin Education in 1996.

8 Gender in disaster beyond men and women

The recent discourse and practice of inclusion in disaster risk reduction have particularly focused on gender. They build on an increasing body of scholarship that has provided evidence for the case. The gendered dimensions of disaster began to garner scholarly attention in the 1980s (Nielsen, 1984; Rivers, 1982; Vaughan, 1987). It is, however, in the 1990s that the study of gender in disaster really took off as a significant field of scholarship (e.g. Anderson, 1994; Dobson, 1994; Enarson, 1998; Fordham, 1998; Forthergill, 1996). Research was then focused on documenting the diversity of people's experiences of what we call disasters and, more particularly, to challenge 'the grounding of disaster theory in men's lives' (Enarson and Phillips, 2008, p. 41).

Such studies found a theoretical home within the so-called vulnerability paradigm, as it had become obvious that natural hazards disproportionally impact women, along with other categories and identities, such as children, older individuals, and people with disabilities, identified along the lines discussed in Chapter 7. Numerous studies have emphasised that, in many places and societies, women are deprived of access to means of protection when facing natural hazards. Yet these are available to men because of unequal power relationships at both local and international levels (e.g. Enarson and Dhar Chakrabarti, 2009; Enarson and Morrow, 1998; Phillips and Morrow, 2008). In that sense, the study of gender in disaster has focused almost exclusively on the particular vulnerabilities as well as capacities of women. Gendered studies of men in disaster are limited (but starting to emerge) although men 'face their own socially constructed roles and expectations which may also place them at risk' (Fordham, 2012, p. 428; see also Enarson and Pease, 2016; Forthergill, 1996; Mishra, 2009).

Strong advocacy by the proponents of the vulnerability paradigm has made gender a major component of the *dispositif* of disaster risk reduction in the 2000s. It has determined that the specific vulnerability of women must be considered in policies and actions aimed at reducing the risk of disaster (Enarson and Fordham, 2001). Further attention has also been given to the role of women in actual disaster risk reduction (Forthergill, 1999). The Sendai Framework for Disaster Risk Reduction (United Nations International Strategy for Disaster Reduction, 2015) states that

DOI: 10.4324/9781315752167-8

women and their participation are critical to effectively managing disaster risk and designing, resourcing and implementing gender-sensitive disaster risk reduction policies, plans and programmes; and adequate capacity building measures need to be taken to empower women for preparedness as well as to build their capacity to secure alternate means of livelihood in post-disaster situations.

As a result, since the 1990s, many international institutions and non-governmental organisations (NGOs) have committed to foster the inclusion of gender – and, more particularly, the participation of women – into their actions towards reducing the risk of disaster (e.g. United Nations International Strategy for Disaster Reduction et al., 2011; OXFAM GB, 2010). These days, challenges remain for embedding gender within government policies of disaster risk reduction. With the exception of a very few countries where timid mentions of women's vulnerabilities and capacities have been made (e.g. in Indonesia and the Philippines), most national legal instruments to reduce the risk of disaster continue to overlook gender.

This encouragement for recognising the importance of women in disaster studies and their inclusion in disaster risk reduction springs mainly from Western (more precisely, European and Northern American) feminist ideas (Bradshaw, 2013, 2014; Fordham, 2012). However, these ideas have been quickly and widely supported by a broad array of scholarship and feminist advocacy movements, which have emerged in Asia and Latin America as part of the growing momentum gained by the vulnerability paradigm (e.g. Begum, 1993; Delica, 1994; Kafi, 1992; Khondker, 1996; Valdés, 1995). Despite a wide range of different approaches among these feminist movements, from liberal to Marxist feminist theories (Enarson and Phillips, 2008), gender in disaster and disaster risk reduction has consistently been conceived from the perspective of the male/man–female/woman binary, based on the material reality of biological differences between sexes, in interaction with the cultural and political forces that shape the roles of both men and women in society.

It is only since the 2010s that a growing number of reports that highlight the experience of people claiming diverse Lesbian, Gay, Bisexual, Transsexual/Transgender, Intersex, Queer and other (LGBTIQ+) identities in disasters have emerged (Dominey-Howes et al., 2014; Gaillard et al., 2017a). These studies are important for drawing attention to the implicit heterosexed gender binary that underpins much disaster risk reduction policy and practice globally, especially those that focus on fostering inclusion. They argue that heteronormative values in both everyday life and during disasters make those who do not identify within the male/man–female/woman binary particularly vulnerable when facing natural hazards. Accessing resources and means of protection designed only for 'men' and 'women' has been shown to cause difficulties and discomforts for various LGBTIQ+ people. For example, in Nepal, people who seek shelter in evacuation centres following flooding have to be recorded as either men or women (Knight and Sollom, 2012) while in the United States of America and Japan, only spouses and husbands are entitled to financial compensation in the event of the loss of a partner (D'Ooge, 2008; Ozawa, 2012). In recent disasters in various countries, these studies have further shown that LGBTIQ+ people are often prevented from accessing evacuation centres, relief goods or counselling services because of the experience of discrimination, stigma and harassment when attempting to do so, such as in Haiti (The International Gay and Lesbian Human Rights Commission, 2011), Canada (Cianfarani, 2013) and Japan (Ozawa, 2012). Ultimately, the experiences of LGBTIQ+ in disaster worsens with the consistent lack of consideration of their concerns in disaster risk reduction policies and practices (Dominey-Howes et al., 2014; Gaillard et al., 2017a). On the other hand, Ong (2017) has shown that the presence of foreign aid workers, especially Westerners who brought their own values to places affected by disasters outside of the West, may provide an opportunity for liberation for many survivors whose LGBTIQ+ identities used to be repressed.

Though certainly welcome and insightful, these emerging LGBTIQ+ studies also prove insufficient for providing a complete and nuanced picture of the gendered dimensions of disaster and hence for contributing to a genuine inclusion agenda. Although these studies are underpinned by a paradigm that attempts to stress diversity and recognise local differences and perspectives, this goal has not yet been fully realised. Pincha and Krishna (2008, pp. 41–42) underlined this gap in their report on the Aravanis of India, affected by the 2004 tsunami, who 'see themselves as neither women nor men' and 'whose gender category cannot be explained using a two-gender framework'. Neither can it be explained or addressed through a particular Western LGBTIQ+ approach, which arguably gives prime importance to sexuality over other roles and identities in society. Gender studies of disaster thus face challenges similar to those of larger studies of disasters in the non-Western world (see Chapter 3). People's experiences of what the West calls disasters cannot always be understood through standard criteria and approaches designed by outsiders (Bhatt, 1998).

It is our intention, in this chapter, to provide an alternative perspective on gender in disaster and disaster risk reduction in order to unpack some of these local nuances. Our argument draws on existing postcolonial studies of gender, an established field of scholarship (e.g. Lazreg, 1994; Lorde, 1984; Lugones, 2007, 2010; Mohanty, 1984, 2003; Oyěwùmí, 1997, 2002; Spivak, 1987, 1993a). Postcolonial studies of gender build upon and contribute to the so-called fields of postmodern feminism (e.g. Butler, 1990, 1995) and include postcolonial queer scholarship (e.g. Boone, 2015; Hawley, 2001; Spurlin, 2006). We believe that this scholarship is particularly helpful in supporting our argument which challenges Western ontologies and their implications in casting a binary view of the world that polarises the West, in its broadest sense, and the rest of the world.

In the following sections, we particularly build upon studies with and stories from people who claim diverse indigenous gender identities in the Philippines, Indonesia and Samoa. We thus draw a hybrid perspective that endeavours to deconstruct dominant Western discourses on gender, including those that have focused on women and men as well as those that have raised the concerns of LGBTIQ+ people. We particularly focus on the nexus between knowledge, norms and power and how this nexus requires considering gender and inclusion from a culturally grounded perspective.

Gender beyond men and women

Our postcolonial exploration of gender in disaster draws upon four key tenets: (1) gender identities are the product of normative expectations in society; (2) gender identities are not monolithic, rigid or stable but diverse, fluid and dynamic; (3) gender identities go beyond the male/man–female/woman binary; and (4) local, indigenous gender identities are very diverse and grounded in local societies and cultures.

It is well accepted that gender identities are social constructs (de Beauvoir, 1949; Rubin, 1975). They mirror the intimate relationship between power and knowledge as expressed in normative discourses that shape our societies. As discussed in previous chapters, discourses, including those which inform our understanding and approach to gender and gender relationships, trickle through all dimensions of society. As such, gender identities and relationships conform to the normative expectations of a dominant discourse on what gender identities are and how they should interact. Identities and interactions are not only about behaviours and social interactions though. They are also about how people conceive and perceive their body. As Canguilhem (1966, p. 135) once said: '*le corps humain est en un sens un produit de l'activite sociale*'[1]. It is through the body that norms and the regulatory injunctions they entail shape people's behaviours so that one's physical appearance, body moves, and way of dressing are defined and controlled within particular discourses. As Butler explains (1990, p. 191):

> gender ought not to be construed as a stable identity or locus of agency from which various acts follow; rather, gender is an identity tenuously constituted in time, instituted in an exterior space through a stylized repetition of acts. The effect of gender is produced through the stylization of the body and, hence, must be understood as the mundane way in which bodily gestures, movements, and styles of various kinds constitute the illusion of an abiding gendered self.

One of the most powerful dimensions of the normative discourses on gender is that they centre upon dimorphism or the biological differences between male and female. Mead (1961, p. 1451) once affirmed that

> all known human societies recognize the anatomic and functional differences between males and females in intricate and complex ways; through insistence on small nuances of behavior in posture, stance,

gait, through language, ornamentation and dress, division of labor, legal social status, religious role, etc. In all known societies sexual dimorphism is treated as a major differentiating factor of any human being, of the same order as difference in age, the other universal of the same kind.

Dimorphism has led both men and women to be defined and identified dialectically. In de Beauvoir's (1949, p. 244) terms: '*la femme est exclusivement définie dans son rapport avec l'homme*'[2]. The materiality of the biological differences between male and female, Mead's (1949) 'two-sex world', thus underpins an ontological binary that polarises men and women as gender identities. It is within this construct that Robin (1975, p. 159) coined the expression the sex/gender system or the 'the set of arrangements by which a society transforms biological sexuality into products of human activity, and in which these transformed sexual needs are satisfied'. Establishing such dialectical relationships between men and women, as pioneered by de Beauvoir (1949), proved to be a prerequisite for uncovering skewed power relations at the origin of gender-based inequalities and discrimination, which most feminist movements aimed to challenge in the first place. Studies of these gender identities as dialectical social constructs, which depend upon the biological dialectic of dimorphism, have been hugely powerful and have served as a springboard for many feminist movements whose achievements in supporting the recognition of the concerns and agency of women have been immense.

Nonetheless, these movements have been criticised for assuming that sex is a material reality rather than a social construct of its own (Butler, 1990, 1995; Wittig, 1992). In this view, '*le genre serait un* contenu, *et le sexe un* contenant'[3] (Delphy, 1991, p. 92). As Butler (1990, pp. 9–10) continues,

> if the immutable character of sex is contested, perhaps this construct called "sex" is as culturally constructed as gender; indeed, perhaps it was always already gender, with the consequence that the distinction between sex and gender turns out to be no distinction at all. It would make no sense, then, to define gender as the cultural interpretation of sex, if sex itself is a gendered category. (...) This production of sex as the prediscursive ought to be understood as the effect of the apparatus of cultural construction designated by gender.

Moreover, the dialectical discourses on gender that draw on dimorphism and the material reality of biological differences between men and women have been challenged for being grounded in the sole experiences of women in the West (Lazreg, 1994; Lugones, 2007; Oyěwùmí, 1997; Verges, 2019). Butler (2004, p. 10) argues that 'terms such as "masculine" and "feminine" are notoriously changeable; there are social histories for each term; their meanings change radically depending upon geopolitical boundaries and cultural constraints on who is imagining whom, and for what purpose'. In the preface of her illuminating book *The Invention of Women*, Oyěwùmí adds that the fundamental category "woman" – which is foundational in Western gender discourses – simply did not exist in Yorubaland prior to its sustained contact with the West. There was no such preexisting group characterized by shared interests, desires, or social position. The cultural logic of Western social categories is based on an ideology of biological determinism: the conception that biology provides the rationale for the organization of the social world. Thus this cultural logic is actually a 'bio-logic'. Social categories like 'woman' are based on body-type and are elaborated in relation to and in opposition to another category: man; the presence or absence of certain organs determines social position.

Such Western discourses on gender have also been criticised for their monolithic views of women's experiences and identities (Lorde, 1984; Mohanty, 1984). Mohanty (1984, p. 333), in particular, argues that the Western and monolithic view of women and their identity prolongs an imperialist legacy where 'colonization almost invariably implies a relation of structural domination, and a suppression – often violent – of the heterogeneity of the subject(s) in question', including 'the production of the "Third World Woman" as a singular monolithic subject'. She eventually contests (p. 338) 'the construction of "Third World Women" as a homogeneous "powerless" group often located as implicit victims of particular socio-economic systems'. Lazreg (1994, pp. 12–13) adds that

> the Third World feminist critique of gender difference acquires a maverick dimension. It is not carried out from within, with a full knowledge and understanding of the history and the dynamics of the institutions it rejects. It unfolds within an external conceptual frame of reference and according to equally external standards. It may provide explanations but little understanding of gender difference. In this sense, it reinforces the existing "meconnaissance" of these societies and constitutes another instance of knowledge as "cutting." Only this time, the cutters are also those who are generally "cut out" of the fellowship of sisterhood.

It has been the first critical contribution of postcolonial gender studies to deconstruct and challenge this discourse and its implication for crafting policies and actions geared towards supporting women beyond the West (Lazreg, 1994; Mohanty, 2003; Oyĕwùmí, 1997; Verges, 2019).

Furthermore, the understandings of such material differences and associated discourse shape and regulate the relationships between people claiming either identity, including the structural relationships informed by sexuality, marriage and kinship (Butler, 1990, 2004; Kessler and McKenna, 1978; Rich, 1980; Wittig, 1992). This ideal representation of sex and gender has been normative and regulatory across Western societies where it has informed social practices as well as the formal or less formal inclusion of gender in the modern art of government that is governmentality. Rich (1980, p. 657) continues:

In Western tradition, one layer – the romantic – asserts that women are inevitably, even if rashly and tragically, drawn to men (…). In the tradition of the social sciences it asserts that primary love between the sexes is "normal", that women need men as social and economic protectors, for adult sexuality, and for psychological completion; that the heterosexually constituted family is the basic social unit; that women who do not attach their primary intensity to men must be, in functional terms, condemned to an even more devastating outsiderhood than their outsiderhood as women.

Rich (1980) called this discourse one of compulsory heterosexuality, where people are expected to conform to such norms associated with male–female sexual and marriage relationships. Rich (1980, p. 646) further emphasises that to acknowledge that

heterosexuality may not be a "preference" at all but something that has had to be imposed, managed, organized, propagandized, and maintained by force, is an immense step to take if you consider yourself freely and "innately" heterosexual. Yet the failure to examine heterosexuality as an institution is like failing to admit that the economic system called capitalism or the caste system of racism is maintained by a variety of forces, including both physical violence and false consciousness.

In this perspective, Butler (1995), following the path paved by Foucault (1976), recast the dominant discourse on sexuality and gender as heterosexual hegemony, or normative heterosexuality, where sex is a linguistic norm but normativity is a regulatory ideal. Those whose identity does not fit the normative and regulatory expectations of this dominant discourse on gender are 'perceived on a scale ranging from deviant to abhorrent, or simply rendered invisible' (Rich, 1980, p. 632), thus crossing the line between the normal and the pathological outlined by Canguilhem (1966). This was evident as early as 1870 in Westphal's pioneering article (p. 107): '*die Erscheinung der conträren Sexualempfindung angeboren als Symptom eines pathologischen Zustaudes auftreten kann*'[4]. The pathological here refers to a form of emotional and sensual deprivation associated with homosexuality (Rich, 1980), with a biological transformation of the body (transsexual and transgender identities) or simply a biological 'anomaly', as in the case of intersex individuals (Besnier and Alexeyeff, 2014; Kessler and McKenna, 1978). In all these forms of 'deviance' or 'maladjustment' (Delphy, 1991), the norm is set either from the perspective of the Western medical tradition or from that of Western society and culture which accepts marriage as the core of kinship and social relationships.

The rise of the LGBTIQ+ movement to claim identities beyond these norms has led to a broadening of the scope of gender identities and an expansion of our understanding of power relations. Yet, in most cases, Butler (1990, 1995) and Wittig (1992) being notable exceptions, these identities and unequal power relations are still considered with reference to the binary through biological appearance or sexual preferences or in relationship to norms such as those of marriage and kinship (Kessler and McKenna, 1978; Rich, 1980). The ontological underpinning is no longer polarised within the male/man–female/woman binary but in a dialectical relationship with the entirety of this binary. As such, the male/man–female/woman binary continues to serve a structuring purpose in the definition of alternative gender identities.

Postcolonial gender studies have challenged this position by looking at gender identities outside of the male/man–female/woman binary as defined in Western traditions. Such a critique starts with revisiting the ontological foundation of gender categories, as drawn from the material reality of biological dimorphism (Oyĕwùmí, 1997). In this perspective, it is essential, and it is our contention hereafter, that gender be dissociated from biological differences. As such, we support Delphy's (1991) and Butler's (1995) thesis that gender, as a normative and regulatory discourse, precedes sex in shaping people's identities. Therefore, gender identities, independent from biology and sex, can be multiple, depending on the norms and values associated with specific discourses in diverse societies.

In the following sections, we provide vignettes of island Southeast Asia and the Pacific, the Austronesian world, which hosts significant minorities who share indigenous gender identities which do not fit the male/man–female/woman binary, nor do they associate with the LGBTIQ+ identities. Examples are the *whakawahine* of Aotearoa, the *fakaleiti* of Tonga, and the *māhū* of Hawaii and Tahiti. Most are biological males who perform different and specific cultural roles in their societies. For those minorities, sex and sexuality matter less than gender in defining their identity (e.g. Besnier, 1994; Besnier and Alexeyeff, 2014; Schmidt, 2003).

We particularly focus on the *bakla* of the Philippines, the *waria* of Indonesia, and the *fa'afafine* of Samoa, groups for which there are enough accounts of how they deal with what we call disasters as recalled by local researchers or people who claim such identities or both (Andag, 2021; de Sagun, 2021; Doron, 2021; Resilience Development Initiative, 2021; Smith, 2014), in tandem with Western perspectives (Gaillard et al., 2017b; McSherry et al., 2015).

Bakla and disasters in the Philippines

Bakla is a colloquial term. It is, in fact, the contraction of *babae* (woman) and *lalaki* (man). It refers primarily to biologically male individuals who claim a feminine identity, which is captured in the more formal Tagalog word '*binabae*' or 'effeminate'. Most dress, put on make-up and do their hair as women. However, *bakla*'s identity is fluid and constantly negotiated by both those who claim such identity and other members of contemporary Philippine society (Manalansan, 2006; Tan, 1995a). The term *bakla* is therefore neither a rigid nor uncontested social category reflecting a unique indigenous gender identity. For example, it is often applied by cisgender men and women to people claiming a gay identity with masculine attributes, although many gay individuals distance themselves from the term *bakla* because of its alleged derogative association with particular occupations, such as beauticians (Benedicto, 2008; Garcia, 2008).

The identities of the *bakla* do not refer to only a particular sexual behaviour; rather, they express their roles within the household and society and an ability to 'swing' between what heterosexual norms consider male and female tasks and responsibilities (Garcia, 2008). In fact, contemporary discrimination against the *bakla* identities can be attributed to the patriarchal and heteronormative nature of Philippine society, which is the visible legacy of its colonial past. Tan (1995b) explains that *bakla* are socially accepted only if they are confined to certain roles and professions, such as beauticians and couturiers. Those who transgress such social norms are at risk of being ostracised and excluded from society. Within the family, young *bakla* are frequently tasked with doing demanding house chores that span the usual responsibilities of both boys (e.g. fishing and fetching firewood and water) and girls (e.g. cleaning the house, doing the laundry and caring for children). Nonetheless, many effeminate *bakla* openly claim – in fact, perform in Butler's terms – their identity and are recognised for their leadership and initiative in collective activities (Tan, 2001). Yet they often suffer from mockery and discrimination, especially in rural areas.

These everyday forms of discrimination that the heteronormative Western colonial heritage imposes upon the *bakla* are reflected during disasters (Gaillard et al., 2017b). In Irosin, a small town located at the southern tip of Luzon, young *bakla* are often asked by their parents to do the dirty chores such as cleaning up the house in the aftermath of recurrent flash floods. In Masantol, located in the delta of the Pampanga River, *bakla* teenagers use their wide range of household skills to look after young children and do the laundry at home while being asked to undertake demanding tasks like fetching water and firewood amid deep flood water after a powerful cyclone in 2011. In Quezon City, Metro Manila, *bakla* reported that some of their fellow youth were left to eat last and least when their households were affected by two back-to-back powerful cyclones in late 2009. In San Julian, a little town nestled on the Pacific shore of the archipelago that suffered the brunt of Typhoon Yolanda in November 2013, *bakla* couples suffered from discrimination from other residents who felt they were undue recipients of aid dedicated to local families. In most accounts (Andag, 2021; Doron, 2021; Gaillard et al., 2017b), *bakla* lament that their specific concerns, which do not match the injunctions of a heteronormative society inherited from colonial times, are never recognised when they have to evacuate in crowded churches or public buildings. They suffer from lack of privacy, and some feel uncomfortable being around either women or men. Their personal grooming concerns are also the object of jokes from men in male bathrooms where they are assigned. Furthermore, they are often the target of sexual harassment and gender discrimination.

The foregoing accounts reveal that the heteronormative nature and expectations of contemporary Philippine society, as a legacy of its colonial past, underpin discrimination and unequal power relations, to the detriment of people claiming identities that do not fit within the male/man–female/woman binary. However, in the meantime, *bakla*, in dealing with what we call hazards and disasters, also display significant resources and skills that do not conform to such norms and injunctions. These resources and skills particularly mirror their ability to transcend heteronormative expectations associated with tasks and responsibilities assumed to be those of either men or women. *Bakla*'s resources and skills also reflect their sense of initiative and leadership. In Irosin, *bakla* spontaneously walk around their neighbourhood to collect relief goods following flash floods. In San Julian, the local *bakla* organisation started to organise similar collection and distribution of relief goods in the aftermath of Typhoon Yolanda. They sought support from international organisations and NGOs that had mushroomed in the region. Since 2013, similar initiatives as well as theatre performances to raise awareness on both the everyday lives of *bakla* and local hazards have been organised in neighbouring islands by the local NGO *Bisdak Pride* (Doron, 2021). In Quezon City, young *bakla* organised larger relief operations following back-to-back cyclones in 2009. They went to request support not only from their neighbours but also from the local chief executive who provided them with relief goods and logistical support. In San Nicolas, a small town located on the shore of the Taal caldera lake, *bakla* similarly volunteered to distribute relief goods secured from the local government when the volcano erupted in January 2020. The tight network of members of their local organisation also provided invaluable moral support in times of hardship (de Sagun, 2021). In evacuation centres throughout the country, *bakla* are often those who spontaneously care for babies and young children, tasks usually assigned to women. Some do the cooking and the cleaning, tasks similarly associated with women.

Although the Philippines is one of the very few countries where recent laws acknowledge the particular concerns of women during disasters, there is no official recognition of gender identities that do not conform with the male/man–female/woman binary. Indeed, the Philippines' 2010 Disaster Risk Reduction and Management Act is framed from the perspective of Western governmentality. The concerns, resources and skills of the *bakla* are systematically overlooked in actions designed to reduce disaster risk because they do not meet the heteronormative expectations of the hegemonic *dispositif* of disaster risk reduction developed from Western discourses on disaster. Nor do the identities of the *bakla* match one of the Western categories/labels that we discussed in Chapter 7. In fact, activities developed by the authorities or NGOs after the injunctions of the 2010 Disaster Risk Reduction and Management Act often prove redundant with what *bakla* spontaneously do to address the risk of disaster. As such, they signal some form of *contre conduites* and resistance that we unpack in Chapter 9.

Waria and disasters in Indonesia

In Indonesia, the dominant gender minority is called *waria*, a contraction of *wanita* (woman) and *pria* (man). *Waria* are biologically male individuals who adopt distinctly feminine features and identity. Yet *waria* identities are subjective and historically constructed and nowadays intersect with Western gay identities (Boellstorff, 2004). They reflect the views of both those claiming such identities and 'outsiders'. In fact, *Waria* are stereotyped as entertainers and sex workers, although many actually work in the beauty industry, operating salons even in remote rural areas. In addition, the prevailing social norms are shaped by the dominant religious view that being 'gay' (in Western terms) and being Muslim are incompatible (Boellstorff, 2005). As a result, if most *waria* dress, put on make-up, do their hair and nowadays even resort to silicon injections to look like women, some also wear men's clothes or mix men's and women's dressing items to avoid being harassed in public spaces (Boellstorff 2004). *Waria* thus does not refer to a homogenous social group, although Kortschak (2010, p. 141) notes that *waria* usually share 'a cohesive subculture with strong social links between members'.

Many *waria* living in Yogyakarta and neighbouring villages on the slopes of Mt Merapi were affected by the eruption of the volcano between October and November 2010. However, most of them were invisible in the numerous evacuation centres set up to shelter those who suffered from the eruption. A *waria* leader remarked that generally *waria* chose not to stay in temporary shelters but rather sought help from and stayed with friends for fear of facing discrimination and hostility in the evacuation sites (Balgos et al., 2012). In fact, this is compounded by their institutional invisibility, in which the heteronormative official guidelines for staff managing evacuation centres recommend that they list evacuees only as women, men, girls or boys, which are the Western categories recommended by international agreements such as the Sendai Framework for Disaster Risk Reduction. In this context, there is no political space for *waria* to claim appropriate recognition of their distinct concerns within a society that imposes heterosexuality as the norm. Although the heteronormative nature of contemporary Indonesian society is not inherited from Western colonialization, it still reflects a foreign heritage because Islam was introduced by Arab traders between the thirteenth and sixteenth centuries (Effendy, 2003; Wanandi, 2002).

Though invisible in the evacuation centres and in official documents, *waria* did not remain passive 'victims' following the eruption of Mt Merapi. Members of People Like Us (PLU), a *waria* NGO, visited the evacuation sites in Magelang, an area that received little assistance from the government and other NGOs. Although they wanted to give money, it was something they did not have. As a result, they decided to do what they knew best, which was providing haircuts and make-up services to the people in evacuation centres. For several days, more than 20 members of PLU took part in the activity. More than 200 evacuees (men, women and children) benefited from the group's free haircuts and make-up services. As such, the action of the *waria*, despite being short-term, addressed an aspect of these human concerns usually overlooked in the plans and guidelines designed after the governmentality of disaster. In fact, evacuees were initially reluctant to let *waria* volunteers into the evacuation centres and even laughed at them. Eventually, though, as one *waria* recalled, 'the group left the evacuation site with the gratitude of the evacuees and the appreciation shown to us provided a sense of fulfilment and hope that people would change their perspective and attitude towards us' (Gaillard et al., 2017b, p. 437). The group eventually collected a considerable amount of money from friends who attended the 'drag' events they organised in Yogyakarta. They were thus able to reach more evacuation sites (Balgos et al., 2012).

Other local *waria* NGOs across the country report similar initiatives (Resilience Development Initiative, 2021). In Jakarta, many members of *Teater Sanggar Seroja* find themselves isolated at home, sick and deprived of their usual modest incomes as buskers when the waters of the Kali Duri flood the streets. As a response, the organisation not only is providing peer support to its member and free meals but also has decided to build its own health centre to serve both the local *waria* and other citizens in the neighbourhood. In Nusa Tenggara Timur, members of the local waria NGO *Fajar Sikka* regularly raise funds to support their members as well as other local residents in times of drought and flooding. They have also used their funds to organise reforestation activities to prevent the occurrence of these phenomena in the first place. In Bali, local *waria* organisations raise funds through creating and selling artworks out of recycled materials. They have organised theatre performances to entertain those affected by disasters and raise awareness. In all instances, local *waria* acknowledge that their experience in dealing with everyday discrimination and hardship helps them overcome the occurrence of floods, droughts, earthquakes and volcanic eruptions.

However, neither the unique concerns nor the significant and specific resources and skills displayed by *waria* in facing what we call disaster are recognised by government agencies and NGOs engaged in disaster risk reduction in Indonesia. For example, the 2007 legal framework for reducing the risk of disaster in the country, designed after the standards and heteronormative injunctions of the Western art of government that is governmentality, does not make any mention of the *waria* whose identities do not meet any of the categories of 'vulnerable groups'. As such, similarly to the case of the *bakla* in neighbouring Philippines, the heteronormative expectations of governmentality and of the broader contemporary Indonesian society underpin skewed power relations between gender identities to the detriment of *waria* who find themselves discriminated both within everyday life and in dealing with what we call disasters. The spontaneous and endogenous responses of the latter to these disasters, which reflect their identities outside of the male/man–female/woman binary, therefore sit beyond the *dispositif* of disaster risk reduction and its heteronormative expectations. As in the Philippines, it signals both the limitations of disaster risk reduction as the Western art of governing disasters and the resistance to this art of government as designed in the West, a form of *contre conduite* in Foucault's terms (Chapters 1 and 9).

Fa'afafine and disasters in Samoa

Fa'afafine literally means 'in the manner of a woman' and refers to biologically male individuals who claim a feminine identity in Samoa. *Fa'afafine* usually dress and put on make-up as women and are an essential dimension of traditional Samoan society. Their ability to perform what Western patriarchal societies typically associate with male and female tasks in everyday life makes them crucial actors in their households and within the broader society (Sua'ali'i, 2001). At the national level, some hold senior positions in government agencies, while the country's prime minister is the patron of the Samoa Fa'afafine Association (SFA). Their leadership in organising social events is, in fact, widely recognised. However, the recent impact of globalisation and the associated importance given to sexuality in defining gender identities in the mind of many Samoans have led *fa'fafine* to actually suffer from 'very real social marginalization', in the words of Schmidt (2003, p. 418). Western influence has also brought changes in *fa'afafine*'s expression of identity; a few have undergone sex-change operations while others now dress as men. As a result, the *fa'afafine* identities nowadays are neither homogenous nor static (Schmidt, 2005).

In times of disaster, it is the ability of *fa'afafine* to undertake a wide range of tasks that proves most significant (Gaillard et al., 2017b; Smith, 2014). During the 2009 tsunami and 2012 Cyclone Evan, both of which severely affected the country, *fa'afafine* were spontaneously at the forefront of rescue operations. For example, Carol, 49 years old, pulled 12 dead bodies from the water in the immediate aftermath of the tsunami; a task usually expected to be performed by men. Others fetched firewood or harvested taros to provide food to their families. At the same time, many *fa'afafine* did household chores that are usually female tasks, such as caring for the babies, cooking, and doing the laundry. These multiple skills acquired in daily life proved essential at both household and village/neighbourhood levels to alleviate hardship associated with the disaster. In addition, unlike other adult members of the households, *fa'afafine* note that they do not have children to look after, which provides them with more time for extra collective and household activities (Smith, 2014). For example, when Cyclone Evan struck, some *fa'afafine* came to help their neighbours, especially children and older people, evacuate their flooded houses. Despite this significant contribution to alleviating the impact of disasters, many *fa'afafine* who had to evacuate in public shelters in the aftermath of Cyclone Evan felt discriminated against. They were particularly uncomfortable using toilet and shower facilities where they felt rejected by both men and women (Smith, 2014).

At the national level, leaders and members of the SFA acknowledge that the organisation was a crucial driver of relief operations conducted in the aftermath of Cyclone Evan (Smith, 2014). *Fa'afafine* came together to collect and distribute relief goods in affected villages. They were supported by a tight network of *fa'afafine* holding positions in government institutions, such as the Ministry of Health, the Ministry of Women and Social

Development, the Ministry of Finance, the Samoa Bureau of Statistics and the Disaster Management Office. The close relationships among *fa'afafine* all over the country and within the government also proved critical in coordinating emergency operations after the 2009 tsunami. *Fa'afafine* holding government positions recognise that friendship and trust were used to overcome lengthy bureaucratic procedures to secure relief goods and other urgent items. For example, sign-off procedures were postponed to quickly secure a water supply or ink for office printers (Gaillard et al., 2017b). These constitute further examples of *contre conduites.*

Fa'afafine further play an active role in everyday initiatives geared towards reducing disaster risk. In 2011, the annual *Miss Fa'afafine* pageant, organised by the SFA, focused on protecting the environment. Some contestants performed acts geared towards raising awareness of what we call natural hazards. The impact of this show is very significant in Samoa and is awaited by thousands of viewers every year. In that way, *fa'afafine* were able to contribute to the governmentality of disaster in a regular cultural event that matters to the larger society (Gaillard et al., 2017b). This ability to both organise and contribute to reducing risk, as framed within and outside of the governmentality of disaster, is not formally recognised in the government's 2009 National Disaster Management Plan. However, since this legal instrument does not discriminate against any gender identities either, *fa'afafine* are regularly invited to contribute to disaster risk reduction activities and discussions. In addition, the SFA is a member of the Samoa Umbrella of NGOs, which sits on the national Disaster Advisory Committee. *Fa'afafine* thus have a wider political space to express their concerns and contribute to disaster risk reduction than *waria* in Indonesia and *bakla* in the Philippines.

Gender minorities in disaster and disaster risk reduction

These empirical vignettes from the Philippines, Indonesia and Samoa teach us two things. First, the Western and heteronormative discourses on gender that dominate the governmentality of disaster, including the particular discourse on inclusion, do not only fail to recognise the unique concerns and abilities of indigenous gender minorities beyond the West; Western and heteronormative norms and values discriminate against them. Second, the experiences of *bakla*, *waria* and *fa'afafine* in what we call disaster challenge our very understanding of gender, articulated within the dominant male/man–female/woman binary and LGBTIQ+ identities, and the scientific narratives that sustain the *dispositif* of disaster risk reduction. We tackle the first of these two issues in this section and the second one in the subsequent section.

Neither biology (i.e. their body) nor their sexual orientation is relevant in explaining *bakla*, *waria* and *fa'afafine*'s experiences in facing what we call disasters. What matters most is their broader, everyday position and role in their household and society. *Bakla*, *waria* and *fa'afafine* are all biologically male individuals who regularly carry out tasks that Western, heteronormative and patriarchal discourses typically associate with women. They are, however, able to undertake responsibilities associated with men when needed, such as in times of disaster, where they shift from one role to another. In that sense, the experiences of *bakla, waria* and *fa'afafine* transcend heteronormative expectations of gendered roles and relationships in society. Therefore, the assignment of such roles and tasks on the basis of identities attached to the material reality of biological differences (i.e. dimorphism) no longer makes sense.

Bakla, waria and *fa'afafine* constitute unique gender groups or minorities in their own societies. Yet the growing influence of Western discourses on gender, especially of those identities that do not fit the male/man–female/ woman binary, has led indigenous gender minorities to be 'framed' through the lenses of homosexuality and transsexuality (Altman, 1996; Besnier and Alexeyeff, 2014). This Westernisation of gender identities that fit neither the male/man–female/woman binary nor any of the LGBTIQ+ identities extends a pattern that both Mohanty (1984) and Spivak (1985) flagged as a form of imperialism, which imposes discourses on gender that are seemingly different when regarded through Western eyes but that all look hegemonic, monolithic and normative when considered from beyond the West.

These discourses are sustained by powerful, seemingly diverse, scientific narratives grounded in Western ontologies and epistemologies that support the project of modernity (Lazreg, 1994; Lugones, 2007; Mohanty, 1984; Oyěwùmí, 1997; Spivak, 1985). In fact, one may argue that the Western discourses on gender have been so successful that they have imposed Western gender categories, positions and roles within society as common sense, in Gramsci's (1971) terms. This common sense is enforced through the modern art of government that is governmentality, or more explicitly here biopolitics, and posits that the body and the biological are at the core of both people's gender identities and the *dispositif* of disaster risk reduction in its normative, regulatory and disciplinary nature (Foucault, 2004a, b).

These discourses on gender are also part of a broader imperialist agenda. As a result, they cannot be dissociated from the broader processes of the Westernisation of the world which have survived the collapse of most colonial empires. Processes that mirror the rolling-out of governmentality throughout the world we unpacked in Chapter 5. As such, in the Philippines, Indonesia and Samoa as well as other places around the world, non-Western gender identities increasingly are associated by outsiders with deviant sexual practices strongly condemned by Christian and Muslim societies, although, as Besnier (1994, p. 300) stated, 'sexual relations with men are seen as an optional consequence of gender liminality, rather than its determiner, prerequisite, or primary attribute'. In this context, indigenous gender minorities such as *bakla, waria* and *fa'afafine* are progressively being discriminated by the rest of the society as well as government policies that are regulatory and normative in nature (Besnier and Alexeyeff, 2014;

Schmidt, 2003). In return, discrimination often leads to hardship in facing natural hazards.

Such discrimination reveals multiple layers of unequal power relations and illustrates some of the points that we made in Chapter 7. If the contribution of feminist and LGBTIQ+ movements in Western societies has been admirable when it comes to uncovering skewed power relations between men, women and people who claim LGBTIQ+ identities, the same movements have also forced the inclusion of people who do not claim similar identities within discourses that do not suit them. As such, these discourses have discriminated against people whose position within society previously never was at the margin. It is therefore the hegemonic and normative nature of the Western discourses on gender centred on categories identified after the male/man–female/woman binary, and articulated around biological dimorphism and sexuality, that should be reconsidered wherever it does not suit local world views/senses, whether these discourses and categories have become common sense or not.

Furthermore, the forced inclusion or invisibility of these indigenous gender minorities in a dominant men/male–women/female framework or through any of the LGBTIQ+ categories for considering and addressing gender in disaster challenges their intrinsic ability to deal with natural hazards and disasters. At worst, they are discriminated against, mocked and deprived of access to resources and means of protection available to men and women. *Waria* of Indonesia lack access to evacuation centres while, in some instances, young *bakla* of the Philippines are deprived of enough food. At 'best', their specific concerns, resources and skills are unrecognised, as in the case of the *bakla* and *fa'afafine* in Samoa. In these three cases, the root determinants of indigenous gender minorities' ability to face natural hazards are grounded both in the changing structure of the society and increasing impact of neoliberalism as the dominant ideology and in a homogenising Western approach to gender in disaster risk reduction. As an example of the latter, the most recent version of the universal Sphere humanitarian standards takes a heteronormative and regulatory approach to providing assistance following disaster that does not consider indigenous gender minorities and continues to specify resources (e.g. toilet facilities) to be provided in times of disaster on the basis of the male/man–female/woman binary (Sphere Association, 2018).

The dominant discourse on gender in disaster constitute a powerful example of understanding and practices designed after Western ontologies and epistemologies to the detriment of local realities. They underpin homogenising and standardised disaster risk reduction policies, firmly embedded within the normative and regulatory *dispositif* that we discussed in Chapter 5, which continue to foster a transfer of knowledge and experience from the West to the rest of the world (Bankoff, 2001; Hewitt, 1983). In many instances, practitioners in regions most affected by disasters have come to include gender programmes as a sign of accountability to Western donors which require 'gender-sensitive' (sic) disaster risk reduction. As an example, the Indonesian National Action Plan for DRR, supported by the World Bank Global Facility for Disaster Reduction and Recovery and the United Nations Development Programme, claims to mainstream gender (Badan Nasional Penanggulangan Bencana, 2010) while in practice *waria* continue to be disregarded if not discriminated against.

Such a normative and regulatory approach of disasters and disaster risk reduction further results in overlooking the intrinsic knowledge, resources and skills of indigenous gender minorities in facing natural hazards and disasters. The experiences of *bakla*, *waria* and *fa'afafine* in disaster mirror initiatives in sharing knowledge, raising support and assisting in evacuation and relief operations and whatever other locally relevant actions that transcend the heteronormative expectations and injunctions of the hegemonic *dispositif* of disaster risk reduction as imposed by the West and supported by regulatory international agreements. Rather, comprehending such contributions of *bakla*, *waria* and *fa'afafine* to reducing the risk of disaster requires a fine-grained knowledge of local contexts and their diversity. This clashes with the hegemonic art of governing disasters that is governmentality as well as the short-term and often rushed practices of disaster risk reduction, as discussed in Chapter 5, especially when the inclusion of gender does not provide much space for considering gender beyond the Western male/ man–female/woman binary, as emphasised in Chapter 7.

Beyond men and women in disaster

The foregoing vignettes documenting the experiences of *bakla*, *waria* and *fa'afafine* in disaster further challenge scientific narratives on gender, and hence gender and disaster, as they have emerged in the West. Considering gender in disaster from a postcolonial lens opens up a range of perspectives that allow us to contest some of the otherwise-granted ontological and epistemological assumptions that underpin disaster studies.

The most obvious of these perspectives is that the postcolonial lens contests the binary nature of gender as framed in the modern *épistémè* of the West. The identities of indigenous gender minorities in island Southeast Asia and the Pacific cannot be seen within the dialectical relationship between men and women as shaped by Western understandings of gender. Nor can they be considered in relationship with the male/man–female/ woman binary in its entirety as per the dominant understanding of LGB-TIQ+ minorities and despite the powerful contribution and relevance of the latter in the West. Indigenous gender identities in island Southeast Asia and the Pacific, as elsewhere in the world, should be seen as standalone identities within specific cultural contexts. These identities are no more homogenous than those of the West, nor are they to be taken in a silo out of other intersecting traits that may be associated with diverse cultural norms and values (Butler, 2004; Oyěwùmí, 1997).

Furthermore, the limits of the male/man–female/woman binary beyond the West reinforce that gender should be seen independently from the materiality of biological dimorphism. The role and position in society of *bakla, waria* and *fa'afafine* result from norms and values set in particular cultures. These norms and values shape their relationships with other people claiming different identities, the tasks and contribution they take on within their households and society as well as their dress code and gestures. Therefore, the relationship to the body and its material reality is a consequence of social and cultural expectations rather than a prerequisite to gender identity (Delphy, 1991; Wittig, 1992). Therefore, as Butler (2004, p. 8) says,

> the critique of gender norms must be situated within the context of lives as they are lived and must be guided by the question of what maximizes the possibilities for a livable life, what minimizes the possibility of an unbearable life, or indeed, social or literal death.

The body and the biological (i.e. dimorphism) therefore lose their underpinning role in shaping gender identities. As per Oyěwùmí's (1997, p. 13) powerful words of introduction to *The Invention of Women*,

> though feminism in origin, by definition, and by practice is a universalizing discourse, the concerns and questions that have informed it are Western (…). As such, feminism remains enframed by the tunnel vision and the bio-logic of other Western discourses.

It is the very existence of the concept of gender that then starts to vacillate. Oyěwùmí (1997, p. 11) adds that

> [t]he potential value of Western feminist social constructionism remains, therefore, largely unfulfilled, because feminism, like most other Western theoretical frameworks for interpreting the social world, cannot get away from the prism of biology that necessarily perceives social hierarchies as natural. Consequently, in cross-cultural gender studies, theorists impose Western categories on non-Western cultures and then project such categories as natural. The way in which dissimilar constructions of the social world in other cultures are used as "evidence" for the constructedness of gender and the insistence that these cross-cultural constructions are gender categories as they operate in the West nullify the alternatives offered by the non-Western cultures and undermine the claim that gender is a social construction.

Gender is another concept of Latin etymology that travelled into English through French around the fourteenth century. Interestingly, its grammatical origin refers to three categories, as per the Latin language. One may argue that it is then logical that, in Western perspectives, those who do not fit the male/man–female/woman binary in non-Western cultures have all been gathered under a third gender category (Herdt, 1994), which mirrors the grammatical neuter category. Yet, if one is to fully step out of the binary and associated Western ontologies and epistemologies, the very existence of such a thing as another, third form of gender becomes irrelevant. Oyěwùmí (1997, pp. 11–12) writes:

> Western ideas are imposed when non-Western social categories are assimilated into the gender framework that emerged from a specific sociohistorical and philosophical tradition. An example is the "discovery" of what has been labeled "third gender" or "alternative genders" in a number of non-Western cultures. The fact that the African "woman marriage", the Native American "berdache", and the South Asian "hijra" are presented as gender categories incorporates them into the Western bio-logic and gendered framework without explication of their own sociocultural histories and constructions. A number of questions are pertinent here. Are these social categories seen as gendered in the cultures in question? From whose perspective are they gendered? In fact, even the appropriateness of naming them "third gender" is questionable since the Western cultural system, which uses biology to map the social world, precludes the possibility of more than two genders because gender is the elaboration of the perceived sexual dimorphism of the human body into the social realm.

We ultimately follow the argument of Oyěwùmí, Butler and Wittig that, in Western thoughts, the body–gender binary entertains a broader dialectic between the biological and identity. This dialectic is also evident in the delineation of other social categories/labels, as per our discussion in Chapter 7, including those associated with race and ethnicity, age, disability and even homelessness and imprisonment, for which perception of the body and biological is essential. Wittig (1980, pp. 76–77) famously wrote that

> en admettant qu'il y a une division "naturelle" entre les femmes et les hommes, nous naturalisons l'histoire, nous faisons comme si les hommes et les femmes avaient toujours existé et existeront pour toujours. Et non seulement nous naturalisons l'histoire, mais aussi par conséquent nous naturalisons les phénomènes sociaux qui manifestent notre oppression, ce qui revient à rendre tout changement impossible.[5]

A postcolonial lens provides us with an opportunity for change by encouraging us to think beyond this binary and to consider identity and, by extension, people's understanding and responses to what we call natural hazards and disasters, as shaped by norms and values in society rather than by the body and the biological. In fact, the very existence of dimorphism, as a biological fact, is now being contested (Delphy, 1991; Hurtig and Pichevin, 1985; Hoquet, 2016). It is therefore our whole approach to gender, but also to race and ethnicity, dis/ability, age, and so on, that needs to be reconsidered, as these categories become irrelevant beyond the West and wherever else the body and the biological are typically associated with an ontological role in defining people's identities.

One can, in fact, take this point further and argue that the biological–identity binary reflects the even broader/deeper dialectic between nature and culture, which, we have seen, has been so essential in shaping disaster studies and disaster risk reduction and which was once put forward as a key factor to understanding the male/man–female/woman binary (Mead, 1949; Ortner, 1972). The significance of the nature/hazard/biological–culture/vulnerability/identity binary therefore appears to crack from multiple perspectives and its relevance in understanding disasters is seriously compromised. It is a crucial contribution of postcolonial studies to uncover such perspectives and to offer alternative pathways to comprehending people's understanding of what we call disaster, as in the case of indigenous gender identities in island Southeast Asia and the Pacific.

Fostering the inclusion of gender in disaster risk reduction

Recognising that gender identities go beyond the Western male/man–female/woman binary and its biological ontological underpinning does have multiple implications for disaster risk reduction, especially for policies and actions designed to foster inclusion, which we discussed in Chapter 7, for one can assume that the Western *dispositif* of disaster risk reduction will remain for some time.

The most important of these implications is obviously that gender cannot be constrained within the dialectical relationship between men and women. Unequal power relations between women and men are indeed essential to understanding people's experiences of what we call disaster wherever these two categories make sense and capture a local cultural reality. However, these categories, identities and associated power relations, as observed in the West, cannot be considered universal. Again, Lazreg (1994), Lugones (2007), Oyěwùmí (1997) and Wittig (1992) have brilliantly demonstrated that these categories are irrelevant in some contexts and that they are 'myths' in Wittig's words, 'fictions' in Lugones's or 'inventions' in Oyěwùmí's. The same demonstration applies to LGBTIQ+ identities. They are valid in some contexts and therefore should be essential components of actions geared towards reducing the risk of disasters where they are relevant. However, they cannot apply anywhere in the world to all individuals who do not claim an identity as man or woman. Any gender identities are associated with unique social and cultural patterns specific to local contexts rather than the universal materiality of biological differences and dimorphism. As such, gender should be considered a context-specific issue that is cultural in essence. As Oyěwùmí (1997, p. 10) once wrote:

> if gender is socially constructed, then gender cannot behave in the same way across time and space. (…) From a cross-cultural perspective, the significance of this observation is that one cannot assume the social organization of one culture (the dominant West included) as universal or the interpretations of the experiences of one culture as explaining another one.

Nor can any of these context-specific gendered categories and identities, including those of men and women, be considered monolithic. They intersect with diverse economic and cultural traits that reflect other dimensions of local societies. This observation entails that no single individual is in position to fully re-present/represent a collective and allegedly homogenous whole. As Spivak (1988, p. 278) summarised in the context of feminist movements: 'nor does the solution lie in the positivist inclusion of a monolithic collectivity of "women" in the list of the oppressed whose unfractured subjectivity allows them to speak for themselves against an equally monolithic "same system"'. It is indeed the essence of inclusion to recognise and consider such diversity, including within the Western *dispositif* of disaster risk reduction.

The evident consequence of the two previous observations is that there cannot be any standard approach to inclusion, nor can there be any magical framework, methodology and toolkit that will work anywhere in the world with any gender minorities. Hence, here lies the relevance of genuine participatory pluralism as discussed in Chapter 4. Initiatives geared towards including indigenous gender minorities in disaster risk reduction have to be embedded within their broader cultural context. They require drawing on local epistemologies and world views/senses, including in considering whether there is such a thing as a disaster as conceptualised in the West and, if appropriate, in defining the contours of alternative/cognate areas of concern. This usually entails that locals and/or those who are to be included, in their diversity, take the lead of an organic process where the 'terms' of inclusion are defined by the locals. As such, this process of inclusion is likely to build on the intrinsic knowledge, resources and skills of indigenous gender minorities, within their specific cultural context, rather than on alleged indicators of vulnerability and capacities as defined after Western scientific paradigms. It is about considering the organic *contre conduites* and tactics of resistance that indigenous gender minorities resort to in facing what we call natural hazards. These *contre conduites* and tactics frequently fall beyond the realm of the governmentality of disaster, as devised by the West (Chapter 9).

This process cannot happen in a silo, as we have seen in the previous chapter. Fostering inclusion is about navigating unequal power relations between gender minorities and people claiming other identities within specific contexts, where such unequal power relations may reflect traditions and deeply entrenched cultural values and norms. Therefore, to foster the inclusion of gender minorities, those with more power should recognise others and otherness and the diverse concerns, knowledge, skills and resources of gender minorities. This process requires trust and hence a dialogue between gender minorities and those who claim different identities. It is about levelling power relations (see Chapter 10).

However, this process cannot be romanticised. It is complex and tangled within an intricate web of structural forces that see local cultural values and norms clashing with the globalising and imperialistic values and norms brought by neoliberalism. The increasing influence of Western LGBTIQ+ movements in the non-Western world is one of these forces that tend to blur the divide between indigenous and Western gender identities, in a pattern that mirrors the hybridisation of knowledge that we discussed in Chapter 7. In Bhabha's (1994, p. 5) words, one is nowadays to acknowledge that 'this interstitial passage between fixed identifications opens up the possibility of a cultural hybridity that entertains difference without an assumed or imposed hierarchy'.

Notes

1 '*The human body is in one sense a product of social activity*'. From the English edition translated by Carolyn R. Fawcett and published by Zone Books in 1991.
2 '*Woman is exclusively defined in her relationship to man*'. From the English edition translated by Constance Borde and Sheila Malovany-Chevallier and published by Vintage Books in 2011.
3 '*Gender as the* content *with sex as the* container'. From the English translation published in Women's Studies International Forum in 1993.
4 '*The manifestation of contrary sexual feeling can appear innately as a symptom of a pathological condition*'. From the English edition translated by Mark Brustman and available from: https://people.well.com/user/aquarius/westphal.htm.
5 '*By admitting that there is a "natural" division between women and men, we naturalize history, we assume that "men" and "women" have always existed and will always exist. Not only do we naturalize history, but also consequently we naturalize the social phenomena which express our oppression, making change impossible*' from the English version published in 1980 in *Feminist Issues* 1(1):103–111.

9 Power and resistance in disaster risk reduction

The modern art of government that is governmentality, in its disciplinary and regulatory nature, inherently entails *contre conduites* and resistance on the side of those whose behaviour is guided and controlled. Foucault (1976, p. 125) himself recognised that '*là où il y a pouvoir, il y a résistance*'[1]. As such, he cleared up early misunderstandings of his theory of power that some believe deprives those who are dominated from any agency (e.g. de Certeau, 1980). In fact, Foucault (1984b, pp. 28–29) once said: 'what I've said does not mean that we are always trapped, but that we are always free'.

In Foucault's (1976, pp. 125–156) view, resistance is not a single and unilateral force that frontally opposes the exercise of power. As he continues, resistance

> n'est jamais en position d'extériorité par rapport au pouvoir. (…) Elles [resistances] sont donc, elle aussi, distribuées de façon irrégulière: les points, les nœuds, les foyers de résistance sont disséminés avec plus ou moins de densité dans le temps et l'espace, dressant parfois des groupes ou des individus de manière définitive, allumant certains points du corps, certains moments de la vie, certains types de comportement.[2]

This approach to resistance, as an inherent and necessary counterbalance of power within the *dispositif* of governmentality, has stimulated much debate. In this chapter, we particularly build upon Scott's (1990) framing of power and resistance, which he conceptualised in dialogue with Foucault's.

Scott (1990) considers that those upon whom power is exercised publicly accept domination to better subvert strategies of power 'offstage', in an opposition that he framed as public versus hidden transcripts. In his own words: 'resistance is virtually always a stratagem deployed by a weaker party in thwarting the claims of an institutional or class opponent who dominates the public exercise of power' (Scott, 1989, p. 52). Such a stratagem is deeply grounded within the everyday lives of those who resist the exercise of power and consists in subverting government and institutions' strategies of control and domination (Scott, 1985, 1990). In sum, Scott, very much like Foucault, approaches resistance as an active response to the exercise of power.

DOI: 10.4324/9781315752167-9

According to Scott, resistance, as an active response to the exercise of power, differs from Gramsci's theory of hegemony within which those who are dominated are believed to lack consciousness to actually resist. We believe that this is a skewed reading of Gramsci, who indeed considered resistance from a defensive, rather than a proactive, perspective that nonetheless requires inventiveness and a tactical approach. This perspective builds upon '*perpetuo fermento*'[3], rather than passiveness, among those who are dominated (Gramsci, 1930). In Gramsci's own words, '*esso* [the war of manoeuvre] *deve considerarsi ridotto a funzione tattica più che strategica*'[4] (Gramsci, 1929–35, Q13, XXX, §24). Tactical resistance that he called war of position.

In this chapter, we draw on Scott's approach to resistance as a convenient means to unpack the *dispositif* of disaster risk reduction. Our intent is to show that, beyond the West, the *dispositif* of disaster risk reduction fails to accomplish its mission to fully control and normalise people's behaviour because the very governmentality of disaster, as the modern art of government of the West, fails to consider the cultural reality of other societies. A postcolonial lens here proves useful to emphasise that there are interstices and cracks in the system through which people design their own tactics of resistance to what, in this book, we call natural hazards. We hope to show that there can indeed be effective and sustainable responses to so-called disasters outside of the formal *dispositif* of disaster risk reduction and that these responses are embedded within local culture and everyday life. Patterns of everyday resistance are essential to fully understanding the nature and impact of the governmentality of disaster as, to quote Foucault (1982, p. 780), 'in order to understand what power relations are about, perhaps we should investigate the forms of resistance and attempts made to dissociate these relations'.

We particularly focus on the unique context of prisons as the ultimate Western institution of power where domination is meant to be brutal and total, where the project of discipline, normalisation and control is at its extreme and resistance is believed to be an impossible occurrence. In the subsequent sections, we show that this intuition is a Western theoretical fantasy and that, in practice, people in prison not only subvert the *dispositif* but indeed collaborate with their guards and the penal institution to design their own response to natural and other hazards as they occur in their everyday lives behind bars. We build our argument upon the particular case from the Philippines, a case that is not unique (e.g. Antillano, 2017; Butler et al., 2018; Darke, 2018; Postema et al., 2017; Skarbek, 2014).

Prisons and power *à l'état nu*

Foucault once said that

> la prison est le seul endroit où le pouvoir peut se manifester à l'état nu
> dans ses dimensions les plus excessives, et se justifier comme pouvoir

moral. (…) sa tyrannie brute apparait alors comme domination sereine du Bien sur le Mal, de l'ordre sur le désordre.[5]

(Foucault and Deleuze, 1972, p. 6)

Prisons are indeed the ultimate expression of the disciplinary and regulatory nature of power constitutive of the modern art of government of the West that is governmentality. Foucault used this case to build his argument in *Surveiller et Punir*.

Modern prisons are designed to isolate those who transgress social norms and values, as expressed in the law, so that they no longer constitute a threat to society. Isolation is further conceived as an opportunity for those who are imprisoned to become better people, away from the sins that have led to their crimes. This is the moral obligation of power, in Foucault's quote. Thus, prisons are designed to rehabilitate people and to redress deviant behaviours. The disciplinary purpose of imprisonment is evident in the strict daily schedule of activities imposed on people in prison to guide their behaviour, redress their attitudes and body gestures, and enhance their ability to stick to the norms and values of society by the end of the time they serve behind bars. Unlike in everyday life outside, as discussed in Chapter 5, this approach to control and discipline people and their behaviour is explicit in prison, which is why Foucault refers to prisons as a place and institution where power shows *à l'état nu*.

The explicit and brutal nature of the power is embedded within the material expression of the *dispositif* designed by the West to address crime through imprisonment, that is the prison as a modern fortress, secluded from the rest of society. Fortresses built after Bentham's panopticon are a form of architecture which allows permanent surveillance and control from a central tower and in which those who are watched are unaware of whether they are actually under the permanent gaze of their guards. As such, the panopticon embodies, within a single piece of architecture, the art of surveillance and control with apparent, though obviously limited, freedom that is governmentality as the expression of liberalism. In fact, one may also see the purpose of imprisonment as being to set people free of their sins so that they can live a better life upon their release. Prisons and imprisonment therefore not only are constitutive of the project of modernity of the West: they are its explicit manifestation. This is true within the West but also elsewhere in the world since the Western approach to imprisonment which we just described has been exported though colonialism and imperialism (Agozino, 2003, 2004; Travers, 2019).

The resistance of people in prison to this extreme form of exercising power has been framed mostly from the perspective of what Sykes (1958) famously called the 'society of captives', or 'inmate society'. Sykes and also Wheeler (1961) a few years later have indeed shown that people in prison are compelled to develop unique tactics to cope, practically and mentally, with the context of deprivation that characterises everyday life behind bars, the 'pains of imprisonment' in Sykes own words. These tactics are forms of resistance, often passive rather than active, that usually entail an 'inmate code' (Schrag, 1954). This 'inmate code' relies on a hierarchy of roles and responsibilities among people in prison. These are the roles and responsibilities that shape and regulate the 'inmate society' (Sykes, 1958).

This so-called deprivation approach to understanding 'inmate society' has been challenged by the proponents of the importation model (Kreager and Kruttschnitt, 2018). The latter consider that resistance to the exercise of power in prison and the emergence of an 'inmate society' are imported from outside and thus mirror values and norms that prevail within the broader society (Irwin and Cressey, 1962). Others have suggested that everyday forms of resistance in prison are highly context-specific and vary from one person in prison to another and from one prison to another (Crewe, 2007, 2012). As Crewe (2007, p. 273) adds,

> [i]n the late-modern prison, power has been designed to individualize prisoners, and there is much of it to wield. It is unsurprising that dissent is largely concealed and individualized, rather than writ large in public discourse. Yet the relative absence of overt, collective resistance should not be interpreted as an indication that the prison's power strategies are entirely successful, that there are no hidden transcripts of discontent, or that all prisoners passively accept the terms of penal power.

Indeed, our contention is that, in some contexts, resistance in prison can take a more (pro)active form as in Darke's (2018) co-governance or Narag and Jones' (2017, 2019) shared governance, when the organisations of people in prison, that are frequently gangs, assume leadership or even collaborate with penal institutions to manage and regulate everyday life in prison. In both the Brazilian (Darke, 2018) and Philippine (Narag and Jones, 2017, 2020) cases, gangs step in to level up a lack of government resources, including personnel, to support daily needs and maintain smooth interpersonal relationships among those who are imprisoned. In these circumstances, the role of gangs in prison is similar to that of civil society organisations which make up for the lack of resources of the government outside in, say, addressing poverty among other issues.

These tactics of resistance have explicitly challenged the relevance of the Western *dispositif* designed to address crime through imprisonment (Agozino, 2003, 2004; Crewe, 2007, 2012). Our intention here is to expand this analysis to the study of what the West calls natural hazards and disasters within the premises of prisons. We argue that the obvious forms of resistance to the everyday hardship associated with imprisonment in the Philippines not only challenge the *dispositif* of disaster risk reduction but also contest the common assumption that the people in prison are vulnerable individuals deprived of agency. Understanding such circumstances, however, requires us to comprehend prisons within their inner and broader cultural environments, which are the cultural dimensions of everyday life within prisons as much as within the broader Philippine society.

Prisons in disasters

First and foremost, we know little about what we usually call disasters in prison (Gaillard and Navizet, 2012). One may argue that this lack of knowledge reflects the downside of Chatterjee's political society and of the labelling game that we discussed in Chapter 7. Few lobby groups support people in prison and their traction on the international scene is limited in comparison with that of non-governmental organisations (NGOs) supporting women and children, for example. In addition, the groups advocating on behalf of people in prison first and foremost focus on the rightful nature of imprisonment and the conditions of detention. As a result, people in prison have not yet been recognised in international agreements such as the Sendai Framework for Disaster Risk Reduction, nor are they included in most national policies geared towards reducing the risk of disaster.

Nonetheless, anecdotal evidence reveals that prisons and people in prison are frequently, though silently, affected by what we call disasters. Prisons and people in prisons have featured in a number of 'high-profile disasters' which captured national or international media attention. These include Hurricane Katrina and its impact on New Orleans prisons in 2005, the earthquake that hit Haiti and its capital city Port au Prince in 2010, and multiple flooding episodes in Southern France (Gaillard and Navizet, 2012). In Western discourse, the suffering of people in prison in disaster has been associated with marginalisation, exploitation and stigma. This view largely reflects the reality of imprisonment in the West, where under-resourced prisons have been 'absorbing' a significant part of the urban poor population (Christie, 1993). Prisons therefore are cradles of poverty that stand at the social, economic, political and spatial edge of society.

In the West, people in prison therefore are perceived as vulnerable and deprived of agency because of the intrinsic nature of their situation behind bars. This view assumes that people in prison are stripped of their prior identity so that they can be rehabilitated and become better individuals. The skills and knowledge they hold outside thus are believed to be of little value inside. Once they have stepped behind the walls of their incarceration facilities, people in prison become dependent on the guards and resources provided by the institution that manages the facility, including in times of disaster. In dealing with natural hazards, this may also entail relying on the external assistance provided by armed forces called upon by the prison authorities to cope with the extra task that constitutes an evacuation, such as in the case of flooding (Gaillard and Navizet, 2012; Le Dé and Gaillard, 2017). In parallel, the very same lack of agency associated with the Western approach to imprisonment, when applied in all its brutality, makes prisoners prone to exploitation. They may be exploited when 'rehabilitation' becomes 'cheap and dangerous labour' wherein they are asked to become front-liners in dealing with bushfire or flooding, as observed in the United States of America (Purdum and Meyer, 2020; Smith, 2016, 2019).

Because people in prison – and, by extension, prisons as an institution – are not recognised in disaster risk reduction policies, the responsibility to deal with hazards and disasters in prison falls upon the penal administrations, at both the national and facility levels. Often, though, the limited resources of prison administrations are logically focused on dealing with more pressing everyday needs, so that responding to natural hazards falls between the cracks and becomes an *ad hoc* issue when they occur. For instance, Human Rights Watch (2005) reported that there was no evacuation plan for the Orleans Parish Prison at the time of hurricane Katrina, 'even though the facility had been evacuated during floods in the 1990s' and even though the United States of America Department of Justice published a *Guide to Preparing for and Responding to Prison Emergencies* only two months before the disaster (Schwartz and Barry, 2005).

Natural hazards are treated as a security issue and their occurrence entails a technocratic response crafted after military principles and geared toward making sure that none of the people in prison tries to break away. As such, the intrinsic nature of the independent mechanisms designed to deal with natural hazards in prison does not differ from the principles that guide the *dispositif* of disaster risk reduction, which we discussed in Chapter 5. It is structured after a top-down chain of command, which includes clear hierarchal roles within and outside the prison, draws on emergency plans, and is focused on battle-like actions to secure the facility and control those who are behind bars (Gaillard and Navizet, 2012; Le Dé and Gaillard, 2017; Smith, 2019). A significant difference, though, is that these actions are usually kept confidential and therefore are shared neither with people in prison nor with local authorities. Similarly, emergency plans do not usually include drills and rehearsal exercises since there is a lack of trust in the ability of those imprisoned to conduct such activities peacefully.

In this chapter, we more particularly focus on Philippine prisons that have had to deal with both small- and large-scale hazards over the past few decades yet with no major casualties.

Prisons in the Philippines

Philippine detention facilities consist of short- and longer-term incarceration centres under the fragmented authority of a range of national and local government agencies (Table 9.1). This chapter focuses only on facilities managed by the Bureau of Jail Management and Penology (BJMP), provincial governments and the Bureau of Corrections (BuCor). These facilities host people awaiting trial and judgement, those who have been sentenced to less than three years (all detained in either BJMP or provincial jails) and convicts who have been sentenced for more than three years (incarcerated in the BuCor prisons and penal farms).

The total population detained in BuCor and BJMP facilities has risen quickly and much faster than the overall Philippine population. It grew from 75,699 people in 2003 to 188,278 in mid-2018. Notably, these figures exclude an additionally large and quickly growing population detained in

Table 9.1 The different types of detention facilities in the Philippines

Temporary detention upon arrest	*Transitional (while awaiting sentence) and short-term (<3 years) detention*	*Long-term (>3 years) detention*
- Lock-ups of the Philippine Drug Enforcement Agency - Lock-ups of the National Bureau of Investigation - Lock-ups of the Bureau of Immigration - Lock-ups of the Armed Forces of the Philippines - Lock-ups of the Philippine National Police	- 464 municipal, city and district jails managed by the Bureau of Jail Management and Penology - 474 municipal jails and the Camp Crame Custodial Center managed by the Philippine National Police - 75 provincial and sub-provincial jails managed by provincial governments	- 7 prisons and penal farms managed by the Bureau of Corrections

Juvenile (<18 years old) detention
- Rehabilitation Centers for the Youth and *Bahay Pagasa* managed by the Department of Social Welfare and Development

provincial jails under the authority of provincial governments. This growth of the population behind bars is a consequence of both increasing arrests and the slow processing of trials. For BJMP and provincial facilities, this leads to an increasing inflow but slow outflow of people. At the same time, only a limited number of new detention facilities are being built. As a result, most Philippine jails and prisons are overcrowded (Figure 9.1). This is especially true for BJMP and provincial jails where the overcrowding rate often exceeds 1,000% and peaks are over 3,000%.

Overcrowding results in challenging sanitary and health conditions (Narag, 2005). In 2013, the annual death rate reached 8‰ among people in jail and prison while not exceeding 5‰ for the whole population. Deaths in detention stem from a wide range of illnesses, including tuberculosis, pneumonia, heart attacks and stroke. These often are preceded by more benign and common problems such as cough, diarrhoea, boils and skin diseases. Illnesses usually break out because of poor sanitary conditions and inadequate food supply. These issues are compounded by chronic poverty among a very large majority of people in prison who were poor upon incarceration and become further impoverished behind bars, especially those who do not receive visits from relatives who could support them (Rueda-Acosta, 2015).

Figure 9.1 Main courtyard of the Quezon City Jail, Philippines, around noontime, January 2016

In facing these challenges, BuCor, BJMP and provincial governments most often are understaffed and short of resources. The combined budget of BuCor and BJMP in 2018 was equivalent to 0.09% of the country's gross domestic product. Financial resources therefore are limited, and the daily budget allocated to each person incarcerated for food is usually PHP50.00 (or US$1.00) while that for medicine is PHP5.00. The ratio of people in jail/prison to staff is also very high (20:1 in BuCor facilities and 15:1 in BJMP jails), and jails and prisons receive relatively limited attention and support from outside stakeholders, with the exception of the International Committee of the Red Cross (ICRC), the Commission on Human Rights of the Philippines and various local NGOs and religious groups. In this context, priorities have to be set, and building new infrastructure and providing food to those behind bars currently take precedence over initiatives towards improving health, sanitation, hygiene, formal and informal education and developing income-generating activities.

In these circumstances, jail and prison officials have no choice but to rely on the people in prison and their organisations, including gangs or *pangkat*, to both organise everyday life in their facilities and sustain their own needs (Candaliza-Gutierrez, 2012; Narag, 2005). Consequently, an informal and unique form of shared governance has emerged in all Philippine jails and prisons as a response to the lack of government and outside resources (Narag and Jones, 2017).

This chapter particularly builds upon stories and testimonies collected in 11 jails and prisons across the country between 2015 and 2016 and discussed in more detail elsewhere (Gaillard et al., 2016b). The collection of these stories and testimonies was notably driven by multiple discussions with Raymund Narag, who spent seven years in jail for a crime that was proven he did not commit and who is now a criminology lecturer in the United States of America. As such, the stories and testimonies presented in this chapter are informed by an insider's view of life in Philippine jails. In particular, the way these stories and testimonies were collected (i.e. through group discussions and face-to-face conversation with people in prison), though inspired by Western research methods, are deemed, in the eyes of an insider, the most appropriate in Philippine jails and prisons (Narag, 2005).

Responses of people in prison to disasters in the Philippines

Although people incarcerated in Philippine jails and prisons suffer from a range of everyday diseases, poverty and rarer natural phenomena, they display a wide range of skills and resources that help them overcome these challenges. These skills and resources range from tightknit social networks to creative decision-making and outside support networks. They are embedded within a hierarchy of roles among people in jail and prison and their organisations. These skills and resources are embedded within multiple imbricated layers of responsibilities and actions that help people in jail and prison to cope with everyday hardship as well as natural hazards. These layers start at the individual level and ultimately encompass the entire jail/prison (Figure 9.2). All together, these layers of responsibilities and actions constitute powerful tactics of resistance.

Layer 1: The individual level

Life in jail and prison is about coping with daily adversities, such as diseases and poverty, which further increase when natural phenomena occur. To deal with everyday hardship as well as with what we call natural hazards and disasters, money is essential. As a result, people in jail and prison try, as much as they can, to save small amounts of cash to cope with any possible adverse events. In fact, wardens recognise that they encourage people in prison to save money, within the limits permitted by the law, although many people in prison manage to acquire more. In parallel, praying constitutes another key coping mechanism often practiced several times a day in dorms or in the jail and prison chapels. Finally, in everyday life as much as in the event of a cell evacuation forced by natural hazards, people try to keep busy by, for example, playing chess and chatting with the people who share the same cell. This is done as frequently as possible in order to cope with boredom and separation from family.

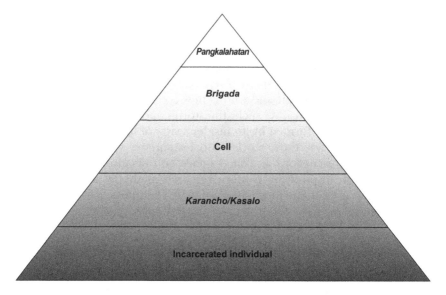

Figure 9.2 The different layers of responsibilities and actions amongst people in Philippine jails and prisons

Layer 2: The **karancho/kasalo** *level*

People in prison recognise that integrating social networks is most often a prerequisite to securing income. The lowest and tightest level of social networks is the *rancho* or *salo*, which refers to a small and tightknit group of individuals who usually share a lack of financial resources and who are tied to a more powerful individual who acts as the patron in a typical client–patron relationship. The clients work (e.g. do the cooking, dishes, laundry and messaging) for their patron in exchange for a weekly wage, extra-financial assistance (e.g. in case of illness or other emergencies) and material support (e.g. soap and toothpaste). In times of hardship, including in dealing with natural hazards, this is the first inter-relational support system mobilised to sustain basic needs. *Rancho/Salo* mates or *karancho/kasalo* share food and soap. More importantly perhaps, a lady incarcerated at the San Mateo Municipal Jail, near Manila, added: 'in times of hardship we talk a lot amongst ourselves. We support each other. This is our own form of counselling'.

Layer 3: The cell level

The informal government structure of the cell provides the upper level of social network. The usually elected cell officers (e.g. *mayor/a* or cell leader, *bise-mayor/a* or associate cell leader, *kulturero/a* or administrative officer, and *bastonero/a* or security officer) look after both the cell's everyday life and unusual situations, including in dealing with natural hazards. For example, *bastonero* of the Quezon City Jail are in charge of fostering readiness (e.g. putting belongings and furniture in safe and higher places) inside the cells whenever there is a typhoon warning. A female *bise-mayora* of the San Mateo Municipal Jail further shared how she and her fellow leaders take the lead in evacuating the facilities in times of flooding, which is a recurrent threat:

> We are organised into four groups of ten people. Each is under the leadership of one cell officer and holds one go box that contains biscuits, candles, medicines and water. Our groups are subdivided through a 'buddy-buddy' system with two people looking after each other: a younger individual and an older one. We (i.e. cell leaders) make sure that the go box are always complete.

Even though all people in prison in every jail are not as well prepared as in San Mateo, stories and testimonies converge in acknowledging that there is usually no privilege or discrimination among individuals when it comes to evacuating their cell in the event of an emergency. A *mayor* from the Pampanga Provincial Jail recalled how they dealt with Typhoon Lando in August 2012:

> When the water started to rise we raised all our belongings on the tarima. We then organised in groups of ten people. We (i.e. cell leaders) took the lead in organising the groups but we were all on the same page once in groups: there were no patron and no client. We were all handcuffed and escorted by the guards along a rope that took us to a new building where we sought shelter.

Leaders also shared that the cell budget, which stems from the contribution of patrons and other sources, is bankrolled to cover regular house chores (e.g. cleaning), unexpected needs (e.g. candles when electricity goes out after a typhoon), maintenance (e.g. roof repair also after a typhoon) and support of the poorest individuals in times of need (e.g. illness and death) in exchange for political allegiance. Financial decision-making revolves around weekly meetings, or *sembol*, involving all people who share the same cell.

Layer 4: The brigada *and* pangkalahatan *levels*

The higher level of social network is the *brigada*. Though often assimilated to gangs or *pangkat*, *brigada* also serve as powerful organisations in many ways similar to people's organisations outside the jails and prisons (Narag and Lee, 2018). *Brigada* are key actors in the overall jail and prison informal government system. They consist of elected officers (e.g. *bosyo* or brigada leader, *bise-bosyo* or brigade associate leader, and *mayor/a*) and rely on financial contributions from each cell affiliated with the *brigada* and other sources, including sales at in-house stores and handicrafts purchased by visitors. Leaders acknowledge that their *brigada* provides support to build infrastructure (e.g. sport grounds and bathrooms) and additional assistance to fellow people in prison in the event of major emergencies or disasters in exchange for political allegiance. An individual from the New Bilibid prison recalled how his *brigada* helped him and his mates overcome the impact of Typhoon Yolanda in November 2013:

> Yolanda wrought havoc here. Many cell roofs and that of the hospital were blown away by the wind. The prison remained flooded for a few days because the drainage system was blocked and we had neither power nor water for one week. We relied on candles and we survived because our brigada purchased gallons of distilled water to support us.

A former *bosyo* of one of the largest *brigada* of the same prison added: 'our brigada organised and paid for the repair of the roofs and water system of the cells we look after'.

The limited number of individuals who choose not to become part of a *brigada* gather in *querna*, which provides the same kind of support but with limited resources because of smaller 'membership', as acknowledged by a *querna* leader from Quezon City Jail. On top of the pyramid, all *brigada* of the same jail/prison collaborate at the *pangkalahatan* level, usually in dialogue with the jail/prison administration. A former warden of the Quezon City Jail discussed how he used to work with the *brigada* leaders when it came to dealing with what we call natural hazards and disasters in his jail despite limited resources:

> Our jail has an OPLAN Lindol (i.e. Operation Plan Earthquake) to respond to an earthquake but it may prove inapplicable (…). We cannot organise drills due to the lack of space but we train the cell leaders during our regular meetings. We tell them that they must make sure that their fellow cellmates drop, cover and hold if the ground starts shaking. They are also in charge of reporting damage to us afterwards.

The outside networks

These layers of coping mechanisms within the jails and prisons are supplemented by outside networks, including relatives and friends at the individual level and NGOs and church groups at the *brigada* level. Relatives and friends bring in additional resources, such as food, medicines and clothes, when dealing with natural hazards. They also provide mental support and comfort so that maintaining contact with their family is one of the main concerns of the respondents in times of hardship. This is significant when the jails are isolated and not accessible because of floods and when people in prison and their families are both coping with the effects of large events such as Typhoon Ondoy in 2009 or Typhoon Yolanda in 2013.

Both people in prison and BJMP officials also emphasise that, in times of disaster, those incarcerated may try to break out of jail or, in exceptional circumstances, are set free on a temporary basis to check on their relatives outside. This happened during and after Typhoon Yolanda. However, most of the 600 individuals who left their jails spontaneously returned after a few days, which makes them eligible for a reduction in sentence should they be convicted. An individual from one jail in Leyte shared what he did after Typhoon Yolanda in November 2013:

> I was very worried about my family as I did not know how bad the damage was in our place. The day after the typhoon I decided to walk home with the consent of the administration. It took me two hours. Everyone was fine and I helped with the initial repair of our house. I stayed for six days. We had food but securing water was difficult so I decided to go back to the jail.

In times of disaster, *brigada* leaders also activate their existing networks of local NGOs and church groups to provide *ad hoc* and usually punctual support, consisting mostly of relief goods and hygiene kits that are distributed to the members of their *brigada*. The same leaders acknowledge that this support is often limited but useful.

Culture, 'inmate society' and disasters

This hierarchical and complex combination of organic responses to what we call natural hazards and disasters, which reflect multiple levels of client–patron relationships, has not emerged spontaneously within the walls of Philippine jails and prisons. Rather, they reflect the broader social and cultural setting wherein jails and prisons sit. The responses of people in prison to natural hazards and disasters reflect key practices, commitments and values that underpin Philippine culture and society. These include the concepts of *kapamaraanan, damayan, pakiramdam* and *bayanihan.*

Kapamaraanan reflects creativity and the ability to cope with adverse situations, including natural hazards and disasters. If *kapamaraanan* sometimes relies on *ad hoc* initiatives and ideas, people in prison also often draw on skills and knowledge acquired prior to imprisonment and imported into the walls of the jails and prisons (Narag, 2005). For example, detained nurses, dentists and doctors make use of their skills inside the jails and prisons, including during times of disaster, to assist their fellow cellmates, often in exchange for financial compensation. Similar creativity and the extension of everyday skills to disaster situations have been documented outside of the prison environment across the Philippine archipelago (Gaillard, 2015; Luna, 2003).

Damayan mirrors compassion and mutual support. It is particularly efficient at the *karancho/kasalo* level where people in prison share resources and support each other in times of hardship, such as when natural hazards strike. This compassion often stems from a sense of shared identity and belonging to a cohort of individuals facing common challenges. In the Philippines, this sense of belonging to a collective of people and the commitment to treat others as one treats oneself is known as *pakikipagkapwa* (Enriquez, 1992; Jocano, 2008). People in prison recognise that *pakikipagkapwa* involves the cultural commitment to get along with their fellow cellmates or *pakikisama* (Jocano, 2008). *Damayan*, as a result of *pakikisama* and *pakikipagakapwa*, provides invaluable support outside of the jails and prisons when people have to cope with food insecurity or spend extended periods of time in crowded evacuation centres in times of floods, cyclones or volcanic eruptions (Bankoff, 2003b; Barrameda and Barrameda, 2011; Tindowen and Bagalayos, 2018).

Pakikipagkapwa also mirrors *pakiramdam*, or sensitivity for others' concerns (Enriquez, 1992). One of its expressions is through *utang na loob*, which roughly translates as 'debt of gratitude'. *Utang na loob*, a controversial concept in Philippine social sciences, ties many Filipinos together through relations of reciprocity (Ileto, 1979; Rafael, 1988; Enriquez, 1992; Jocano, 2008). On the one hand, *pakiramdam* supports horizontal relationships between fellow cellmates whose mutual support through *damayan* often entails reciprocity, as observed outside of jails and prisons (Dalisay, 2008; Gaillard, 2015). On the other hand, reciprocity also underpins the prevalent client–patron relationships that shape the political economy of prisons and broader Philippine society (Hollnsteiner, 1963). It is mirrored in the allegiance and support observed between patrons and clients as well as through the upper assistance provided at the cell and *brigada* levels. This also happens in times of hardship. *Utang na loob* therefore contributes to the hierarchical nature of 'inmate society' that prevails in Philippine jails and prisons.

Bayanihan is also associated with *damayan*, *pakikipagkapwa* and *pakiramdam* (Enriquez, 1992; Jocano, 2008). It reflects the ability of people in prison to come together and cooperate to complete a common task, such as putting out a fire or repairing any facilities damaged by a cyclone. As demonstrated outside of jails and prisons (Bankoff, 2003b; Barrameda and Barrameda, 2011; Luna, 2003), *bayanihan* often is mobilised in times of disaster when people, including those in prison, are confronted with tasks that require significant effort and staff power. In jails and prisons, *bayanihan* is supported by the cell and *brigada* social and political systems. These offer an institutional umbrella for the speedy organisation of *bayanihan* initiatives.

These practices, commitments and values constitute the core of Philippine 'inmate society'. They contribute to binding people in prison together through multiple forms of interactions that reflect compassion, collaboration and reciprocity but also economic and political allegiance at the multiple levels of relationship that was discussed in the previous section. At the New Bilibid Prison, these relationships also encompass ethnicity, which is the foundation of some *brigada* (e.g. *Batang Samar Leyte*, whose members are *Waray,* and *Batang Ilokano,* whose members are *Ilokano*). Leaders among people incarcerated at the New Bilibid Prison say that the practices and values of these *brigada* reflect their own ethnic identity. As such, the 'inmate society' mimics most practices and values as well as interpersonal relationships and the roles of social and economic actors that underpin the structure of the broader Philippine society. Hence, patterns of response to what we call natural hazards and disasters in jails and prisons are unsurprisingly similar to those observed outside (Banzon Bautista, 1993; Bankoff, 2003b; Gaillard, 2015; Tatel, 2011).

Most of these practices, values and roles in the 'inmate society' are formalised in an 'inmate code'. Leaders among people in jail and prison recognise that the details of the code differ slightly across gangs/*pangkat* and jails/prisons but the overall architecture remains similar. The code relies on a series of dos and don'ts that appear in each and every cell of all jails and prisons (Figure 9.3). Although none of the codes we have seen explicitly refers to natural hazards, the duties and commitments listed extend to times of disaster, according to people in prison and their leaders. People in prison who fail to comply with the code, including in times of disaster, appear before a judge and his advisor(s) who pronounce(s) a sentence (usually a form of physical punishment) that is applied by the *bastonero/a.* Thus, the ability to deal with everyday hardship and natural hazards is formalised in an integrated way, which is seldom observed outside of the jails and prisons themselves.

If imprisonment reflects deprivation and social compression (Crewe, 2012), stories from the Philippine prisons, including those by Narag (2005), Narag and Jones (2017, 2020), and Narag and Lee (2018), show that this deprivation and social compression also stimulate coping mechanisms and creativity so that people can survive an extremely harsh everyday environment. They devise social, economic and political tactics to overcome overcrowding, boredom, insalubrity, poverty and political neglect (Narag, 2005). All of these issues are compounded in times of disaster, so that the tactics geared towards everyday hardship extend to dealing with natural hazards. Therefore, the responses of people in jail and prison to natural hazards and disasters do not constitute *extra*-ordinary initiatives but rather reflect everyday practices, roles and values.

In fact, the everyday tactics devised by people in prison to deal with natural hazards and disasters do not emerge spontaneously and are not unique to any specific jail or prison. They stem from pre-existing practices and values imported into jails and prisons (Irwin and Cressey, 1962). They more broadly reflect the nature of Philippine culture and society (Enriquez, 1992; Jocano, 2008). Practices and values, including *kapamaraanan, damayan, pakiramdam* and *bayanihan,* are imported into jails and prisons and are firmly embedded within the 'inmate code' that also relies on a social hierarchy of roles and responsibilities which mirror the social status of people in jail and prison before their incarceration (Clemer, 1940). They constitute powerful tactics of resistance.

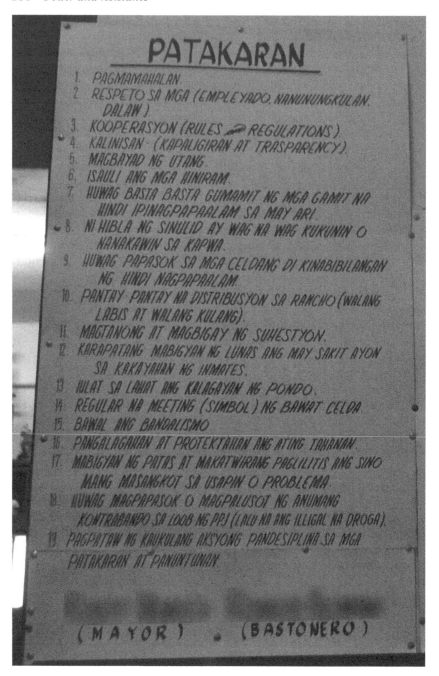

Figure 9.3 Cell rules set by a *pangkat* at the Pampanga Provincial Jail in San Fernando, The Philippines, January 2016

On resistance, disaster and disaster risk reduction

These stories from Philippine jails and prisons reveal that resistance to both the governmentality of disaster and imprisonment is subversive and tactical rather than conflictual and frontal. In all their despair and amid extremely harsh everyday conditions of living, people in prisons build upon opportunities and loopholes in the penal system, including its lack of resources, to deal with natural hazards and disasters on their own and outside of the formal emergency management system set up by the administration or, in some instances, in collaboration with it. They build upon their own resources and skills, imported from the outside and embedded within both Philippine cultural values and the prison structure. This perspective mirrors de Certeau's (1980, pp. XXVII– XXVIII) insightful assessment of Spanish colonisation in what is now Latin America:

> Il y a longtemps qu'on a étudié, par exemple, quelle équivoque lézardait de l'intérieur la "réussite" des colonisateurs espagnols auprès des ethnies indiennes: soumis et même consentants, souvent ces Indiens faisaient des actions rituelles, des représentations ou des lois qui leur étaient imposées autre chose que ce que le conquérant croyait obtenir par elles; ils les subvertissaient non en les rejetant ou en les changeant, mais par leur manière de les utiliser à des fins et en fonction de références étrangères au système qu'ils ne pouvaient fuir. Ils étaient autres, à l'intérieur même de la colonisation qui les "assimilait" extérieurement; leur usage de l'ordre dominant jouait son pouvoir, qu'ils n'avaient pas les moyens de récuser; ils lui échappaient sans le quitter[6].

In Scott's (1989, 1990) terms, actions to respond to hazards and disasters in prison are geared to secure *de facto* gains. These *de facto* gains focus on avoiding or reducing harm. In contrast, actions to deal with hazards and disasters in prison do not aim at securing *de jure* recognition of the nature of the responses or of their success (or failure). In fact, in Philippine jails and prisons, seeking recognition of these responses (e.g. that some incarcerated people hold the key of their cell and therefore can organise evacuation by themselves) may endanger the sustainability of these actions. As a result, everyday forms of resistance in dealing with natural hazards and disasters, which are actions that fall beyond the formal and institutionalised realm of the governmentality of disaster, rarely if ever make the headlines.

Unlike the successful outcomes associated with formal risk reduction or disaster response mechanisms, informal forms of responses remain hidden transcripts, to use Scott's famous term. For example, the local government of the small town of San Francisco in the central Philippines was internationally acclaimed for having activated a proactive response mechanism that allowed all local people to survive Typhoon Yolanda in November 2013. At the same time, the largely informal response of the Bureau of Jail

Management and Penology and other jail authorities, in close collaboration with people in prison, which similarly led to no casualties across the
jails of the entire country, was completely unnoticed despite its impressive
success.

It was not in the interest of the jail administration to recognise that they
had collaborated with people in jail and prison, nor for the prisoners themselves to acknowledge that they hold more power than allowed by the law,
to disclose the reasons behind the success of the Yolanda story. As Scott
(1989, p. 49) summarises:

> it is also […] rare that officials of the state wish to publicize the insub-
> ordination behind everyday resistance. To do so would be to admit that
> their policy is unpopular and, above all, to expose the tenuousness of
> their authority. […] The nature of the acts themselves and the self-in-
> terested muteness of the antagonists thus conspire to create a kind
> of complicitous silence which may all but expunge everyday forms of
> resistance from the official record.

As such, these patterns of resistance that materialise in people's organic
responses to natural hazards and disasters form a subculture, where the
prefix 'sub' stands both for a specific set of cultural traits within a broader
culture but also for the hidden nature of these responses. Subculture is, in
fact, not a new research topic in disaster studies (Anderson, 1965; Wenger
and Weller, 1973; Wenger, 1978), and we agree with Anderson's (1965, p. 3)
definition that refers to 'those subcultural patterns operative in a given area
which are geared towards the solution of problems, both social and non-so-
cial, arising from the awareness of some form of almost periodic disaster
threat'. However, we aim here to offer a political reading of subcultures of
disasters, the hidden side of the prefix 'sub', in that they constitute a form
of resistance to dominant Western discourses and to the governmentality of
disaster they underpin.

In fact, discourses on disaster shaped by Western scientific narratives and
concepts definitely trickle down to Philippine jail guards and wardens as
well as local government officials. However, they do not have much influence on people in prison whose responses to natural hazards happen
independently from their labelling as vulnerable, which is the dominant
narrative on people at the so-called margins of society (Chapter 6). In sum,
the hegemony of Western knowledge, at the core of the argument put forward in this book, affects only the traditional intellectuals, in Gramsci's
term, who hold power in and influence the governmentality of disaster as
designed in the West. Those who actually face what we call natural hazards,
and who may or may not call the same natural phenomena 'hazards', deal
with these very phenomena from their own perspectives.

It is our contention that such patterns of response occur on a wider scale,
beyond prisons, in the Philippines and elsewhere and that a large number
of people, surely not all, tacitly and politely acknowledge and approve the

formal disaster risk reduction initiatives of governments and NGOs but ignore them, as well as the concepts, narratives and discourses that underpin them, when it comes to actually saving their lives and responding to what we call disasters. We assume that this is true for both the dominant technocratic and top-down actions planned by government agencies and the so-called community-based/led/managed initiatives of NGOs. As in Philippine prisons between *pangkat* and guards/wardens, this 'tactical wisdom', to use Scott's words, allows those who are dominated to secure political allegiance and to leverage support for diverse purposes in other instances. 'Everyday resistance, then, by not openly contesting the dominant norms of law, custom, politeness, deference, loyalty and so on leaves the dominant in command of the public stage' (Scott, 1989, p. 57). This hidden transcript of resistance and response to natural hazards and disasters thus remains informal yet often collective and surely political. As Scott (1989, p. 52) continues:

> No formal organizations are created because none are required; and yet a form of coordination is achieved which alerts us that what is happening is by no means merely random individual action. Nor is it too much to suggest that the historical experience of the peasantry has favored such forms of social action because they are opaque to outside surveillance and control.

As such, despite its apparent far-reaching and exhaustive *quadrillage* of society, the governmentality of disaster may partly fail to achieve its goal of controlling and disciplining people and their behaviours when dealing with natural hazards, especially beyond the Western world where we have, in the first place, challenged the very relevance of governmentality as the modern art of government. It is our contention that, in many places around the world, allegedly comprehensive risk assessments, shiny risk reduction plans, and normative drills to rehearse responses to actual disasters are mere mechanisms of upward accountability to authorities of all sorts, such as local and central governments, NGOs and donor agencies. These mechanisms of upward accountability are inherently embedded within the *dispositif* of disaster risk reduction and, as a result, legitimise and self-sustain governmentality as the only way of dealing with what we call disasters as underpinned by the hegemonic *métarécits* of the West.

Everyday forms of resistance to the governmentality of disaster as the hegemonic Western art of government are therefore an intrinsic, though hidden, dimension of what we call disasters in the West. As Foucault suggested,

> if there was no resistance, there would be no power relations. Because it would simply be a matter of obedience. You have to use power relations to refer to the situation where you're not doing what you want. So resistance comes first, and resistance remains superior to the forces

of the process; power relations are obliged to change with the resistance. So I think that resistance is the main word, the key word, in this dynamic.

(Foucault, 1984b, p. 29)

Beyond Western governmentality and disaster risk reduction

Our intention in this chapter is to suggest that more research is needed to validate what is presented here as a hypothesis rather than a definite claim. That the intrinsic disciplinary and regulatory nature of the governmentality of disaster and its apparatus of surveillance and control favours the emergence of informal and possibly widely successful if unrecognised responses to what the West calls natural hazards and disasters. Responses that, in the image of Philippine prisons, are deeply embedded and implicit within their social and cultural context and, as such, form a subculture.

Considering such forms of response to natural hazards as everyday resistance to the governmentality of disaster ultimately allows us to move beyond a dialectical understanding of power relations that oppose the powerful, including both individuals and institutions, and the powerless, including the so-called 'marginalised' social groups, or as Foucault (1976, p. 124) famously said:

> il n'y a pas, au principe des relations de pouvoir, et comme matrice générale, une opposition binaire et globale entre les dominateurs et les dominés, cette dualité se répercutant de haut en bas, et sur des groupes de plus en plus restreints jusque dans les profondeurs du corps social[7].

This approach to resistance and power relations allows us to step outside of the reductionist view of power relations as polarised between a centre and its margins. Responses to natural hazards and disasters, as patterns of everyday resistance, therefore need to be considered in a diffuse and fluid perspective that transcends the centre–margin dialectic. In Foucault's (1976, p. 127) terms:

> on a affaire le plus souvent à des points de résistance mobiles et transitoires, introduisant dans une société des clivages qui se déplacent, brisant des unités et suscitant des regroupements, sillonnant les individus eux-mêmes, les découpant et les remodelant, traçant en eux, dans leur corps et dans leur âme, des régions irréductibles[8].

As such, we suggest going beyond 'researching' people who fall between the cracks of the governmentality of disaster, including prisoners, for the sake of highlighting their concerns on the political scene and lobbying for their (token) inclusion in a form of government, which is governmentality as designed by the West and which inherently fails to provide any space for genuine inclusion.

Rather, challenging the dominant Western discourses on disaster and the governmentality of disaster requires the study and consideration of the specific cultural contexts wherein people, in all their diversity, including prisoners, live and respond to what we call natural hazards and disasters if this is ever relevant in their own eyes. As Gramsci (1971) suggested, patterns of resistance are found in folklore, music, literature and other aspects of popular culture. It is therefore those actually resisting the governmentality of disaster – the emerging organic intellectuals, to stick to Gramsci's lexis – who likely have the best understanding of the culturally embedded tactics devised to deal with natural hazards and disasters.

Studying forms of everyday resistance in disaster risk reduction will be possible only through sneaking through the interstices of both the dominant and dialectical understanding of power relations the polarised view of disaster within the nature/hazard–culture/vulnerability binary. This will be possible only by studying the hidden transcripts of power relations and people's everyday lives and struggles. Studies from below and grounded because, as Foucault (1976, p. 124) once famously stated, '*le pouvoir vient d'en bas*'[9].

Notes

1 '*Where there is power, there is resistance*'. From the English edition translated by Robert Hurley and published by Pantheon Books in 1978.

2 '*is never in a position of exteriority in relation to power.(...) They [resistances] too are distributed in irregular fashion: the points, knots, or focuses of resistance are spread over time and space at varying densities*', at times mobilizing groups or individuals in a definitive way, inflaming certain points of the body, certain moments in life, certain types of behavior'. From the English edition translated by Robert Hurley and published by Pantheon Books in 1978.

3 '*Perpetual ferment*'. From the English edition translated by Quintin Hoare and published by Lawrence & Wishart in 1977.

4 '*The war of manoeuvre must be considered as reduced to more of a tactical than a strategic function*'. From the English edition translated by Quintin Hoare and Geoffrey Nowell Smith and published by International Publishers in 1971.

5 '*Prison is the only place where power is manifested in its naked state, in its most excessive form, and where it is justified as moral force. (...) Its brutal tyranny consequently appears as the serene domination of Good over Evil, of order over disorder*'. From the English translation by Sherry Simon published in *Language, Counter-Memory, Practice: Selected Essays and Interviews by Michel Foucault* by Cornell University Press in 1980.

6 '*The ambiguity that subverted from within the Spanish colonizers' "success" in imposing their own culture on the indigenous Indians is well known. Submissive, and even consenting to their subjection, the Indians nevertheless often made of the rituals, representations, and laws imposed on them something quite different from what their conquerors had in mind; they subverted them not by rejecting or altering them, but by using them with respect to ends and references foreign to the system they had no choice but to accept. They were other within the very colonization that outwardly assimilated them; their use of the dominant social order deflected its power, which they lacked the means to challenge; they escaped it without leaving it*'. From the English edition translated by Steven Rendall and published by the University of California Press in 1984.

7 '*There is no binary and all-encompassing opposition between rulers and ruled at the root of power relations, and serving as a general matrix – no such duality extending from the top down and reacting on more and more limited groups to the very depths of the social body*'. From the English edition translated by Robert Hurley and published by Pantheon Books in 1978.

8 '*More often one is dealing with mobile and transitory points of resistance, producing cleavages in a society that shift about, fracturing unities and effecting regroupings, furrowing across individuals themselves, cutting them up and remolding them, marking off irreducible regions in them, in their bodies and minds*'. From the English edition translated by Robert Hurley and published by Pantheon Books in 1978.

9 '*Power comes from below*'. From the English edition translated by Robert Hurley and published by Pantheon Books in 1978.

10 The invention of disaster

So far throughout this book, we have assumed that there is such a thing as a 'disaster'. This assumption was evident by asking what a disaster actually is when we opened the book (Chapter 1). Both this assumption and this interrogation were essential to locate our endeavour of deconstruction within the scientific narratives that underpin the concept and the broader discourses associated with the governmentality of disaster.

This endeavour of deconstruction from within has shown that what we usually mean by disaster, how we frame the concept (Chapter 2), how we conduct research (Chapters 3 and 4) and how we ultimately deal with disasters on the ground through disaster risk reduction (Chapters 5, 6 and 7) reflect a Eurocentric/Western perspective. There are obviously significant nuances in scientific paradigms and approaches to policy and actions to reduce disaster risk but they are ultimately all firmly grounded within Western ontologies and epistemologies and within the Western art of government that is governmentality.

The hegemony of Western science is particularly evident in the broad acceptance that disasters sit at the interface between nature and culture, often captured through the famous mnemonic 'disaster = hazard × vulnerability' or any of its multiple iterations. This ontological assumption, inherited from Europe's eighteenth century of Enlightenment, has polarised research, most often driven by researchers based in the West, who, in turn, have focused on nature/hazard and culture/vulnerability, sometimes on both sides of the binary (Chapters 3 and 4). Methods to uncover and assess disaster risk have been epistemologically aligned with the Western heritage of the Enlightenment, successively fostering econometric and reductionist approaches, anthropological and particularistic techniques, and pluralistic and participatory initiatives (Chapter 4).

Europe's colonial and imperialist legacy has led such understandings of disaster to dominate and inform an overall monolithic approach to disaster in spite of the seemingly significant, but ultimately marginal, nuances that the different discourses have forged. As a result, it is no surprise that policies and actions geared to reduce disaster risk and alleviate suffering in the time of, or following, what we call disaster have drawn on a *dispositif* that sustains governmentality as the art of government of the West (Chapter 5). The

DOI: 10.4324/9781315752167-10

governmentality of disaster appears to nicely support the goal of the project of modernity, which is to free people from the threat of nature while ensuring some level of control and discipline. Nowadays, this project of modernity is epitomised by the dominant neoliberal discourse.

Our argument throughout this book has been that the hegemony of Western science and its influence on policies and actions to reduce disaster risk have prevented other perspectives from flourishing, especially beyond the West. In this conclusion, it is our intention to briefly outline a research agenda that would allow for these other perspectives to emerge. However, for such an agenda to emerge, one must step outside of the ontological assumption that disasters sit at the interface between nature/hazard–culture/vulnerability and then one must reconsider the very nature of disaster. In fact, one must question whether there is actually such a universal thing as a disaster because, as Ferguson (1993, p. 7) brilliantly summarised,

> [t]he questions we can ask about the world are enabled, and other questions disabled, by the frame that orders the questioning. When we are busy arguing about the questions that appear within a certain frame, the frame itself becomes invisible; we become enframed within it.

The invention of disaster

Is there such a thing as a disaster? The answer is inherently subjective and contextual and it will be up to whoever dares to take our agenda forward to try to answer the question in their own unique context. Our contention, though, is that there may be no easy answer for the very reason that disaster, like any other concept in the Western world view, is an invention. According to Derrida, who helped us frame our endeavour of deconstruction in the first place, the very existence of the concept of disaster depends on our understanding of countless cognate concepts. In Derrida's (1972, p. 11) own words:

> le concept signifié n'est jamais présent en lui-même, dans une présence suffisante qui ne renverrait qu'à elle-même. Tout concept est en droit et essentiellement inscrit dans une chaîne ou dans un système à l'intérieur duquel il renvoie à l'autre, aux autres concepts, par jeu systématique de différences. Un tel jeu, la différance, n'est plus alors simplement un concept mais la possibilité de la conceptualité, du procès et du système conceptuels en général[1].

Each of these different concepts (the *différence*) associated with disaster is similarly dependent on a number of other concepts, and so on, so that in the end the meaning of disaster is deferred (the *différance*) to our understanding of countless other concepts, which often are relative in time and space. In summary, disaster, as much as any other concept, at least in Western languages, does not reflect any transcendental signifier, to use

Derrida's (1967a) term, and is inherently relative. There is therefore no such thing as a disaster per se as '*il n'y a rien hors-texte*'[2] (Derrida, 1967a, p. 127). The meaning of the concept is necessarily mediated by our interpretation of other concepts.

In disaster studies, all of these different and deferring concepts – vulnerability, capacities and resilience – and all the subsequent concepts that allow us to understand the former have a common Western heritage, which is the logocentric heritage of Latin (and Greek) language(s). Yet *différence* and *différance* are both synchronic and diachronic processes. Indeed, as Gramsci (1929–35, Q11, XVIII, §24) once stated, '*nessuno oggi pensa che la parola "dis-astro" sia legata all'astrologia e si ritiene indotto in errore sulle opinioni di chi la usa*'[3]. It is because our contemporary understanding of disaster as a concept reflects one specific tradition and one set of particular meanings associated with cognate concepts, both of which are inherited from the Enlightenment and the *épistémè* of the West. As a result, our interpretation of disaster inherently reflects a Western interpretation of the world. Disaster is therefore a Western invention.

This invention may or may not make sense outside of where all the different and deferring concepts originate: the West. This cognate, though an absolutely essential question, remains largely unanswered because of the hegemonic and normalising nature of Western science in disaster studies. In Clifford's (1986b, p. 119) terms, in the specific context of ethnographic writing: 'the notion that writing is a corruption, that something irretrievably pure is lost when a cultural world is textualized is, after Derrida, seen to be a pervasive, contestable, Western allegory'. This allegory, in literary and ethnographic terms, or invention in our broader perspective needs to be recognised and, as such, poses inherent political and ethical questions. This invention needs to be challenged from within, which is what we have tried to do in this book. It also needs to be subverted by diverse other and standalone perspectives that reflect the multiple realities of world views/senses around the world because, as Mbembe (2013, pp. 26–27) reflected,

> dans la façon de penser, de classifier et d'imaginer les mondes lointains, le discours européen aussi bien savant que populaire a souvent eu recours aux procédures de la fabulation. En présentant comme réels, certains et exacts des faits souvent inventés, il a esquivé la chose qu'il prétendait saisir et a maintenu avec celle-ci un rapport fondamentalement imaginaire, au moment même où il prétendait développer à son sujet des connaissances destinées à en rendre objectivement compte.

Before suggesting how to facilitate such subversion, we propose to briefly detail how disaster as a Western invention that stands on evident shaky grounds has become a solid and durable object of scientific inquiry, shaped a whole field of scholarship, and supported an entire *dispositif* of governmentality. It is particularly crucial to focus on the specific conditions of its emergence, legitimation and institutionalisation (Derrida, 1987). Understanding this fascinating process is an important step towards envisioning not only how other views and narratives may be seen and heard but also how they may subvert the current discourses on disaster and stand beside it, wherever these alternative discourses may be relevant.

From concept to object: the legitimisation of the invention

Because disaster as a concept is inherently a Western invention, the construction of disaster as a scientific and policy object has to build upon fabricated facts. These facts give existence to a concept that has no transcendental reality. This process is not specific to disaster studies and the governmentality of disaster. Social studies of science, in the path of scholars such as Latour (1987) and Latour and Woolgar (1979), have extensively unpacked the construction of scientific objects and their influence on policy. Castree (2014, p. 6) has more specifically unpacked how nature is 'a particularly powerful fiction: it's something made, and no less influential for being an artefact'. It is not our intention here to revisit all dimensions of this process of invention in the context of disaster studies. Rather, we focus on a few significant aspects that allow us to pull together some of the points made earlier in this book.

Facts that contribute to the existence of disaster as a scientific object are fundamentally grounded in the ontological assumption of the West that disasters sit at the interface between nature and culture, when natural hazards, in the specific context of our book, negatively impact a vulnerable society. However, for this ontological assumption to transform into an object that both scientists and governmentality can handle, there needs to be a normative delineation of scope that, in dealing with disasters, always mirrors the scale of the impact. Much has been discussed with regard to when the impact is significant enough to be labelled disaster (e.g. Quarantelli, 1998; Perry and Quarantelli, 2005) and we do not endeavour to revisit this debate here.

Rather, our interest centres on the normative nature of the process and how it reflects Western expectations. In this perspective, we heavily draw on Canguilhem's (1966) seminal exploration of the threshold(s) between the normal and the pathological in medicine. In fact, in English, *dis*-ease and *dis*-aster share the Latin prefix *dis*-, which reflects the negative dimensions of both concepts, a divergence from a normal status, whether it is health (in the case of disease) or everyday life at large (in the case of disaster). This divergence from the normal, the regular and the everyday, as we saw in Chapter 1, has underpinned the long-dominant hazard paradigm in disaster studies.

Indeed, the very concept of *dis*-aster is inherently associated with the delineation of thresholds between the normal and the too much (suffering, damage, impact, etc.). Such thresholds are inherently normative whether they are qualitative or quantitative. Qualitative thresholds refer to the level of disruption that natural hazards cause to the social fabric (as approached by the long and strong sociological tradition of disaster studies) and the ability to cope with such disruption (as defined by many organisations that, as such, justify their 'intervention' in responding to disasters). In both cases, the thresholds are set against normative expectations of what 'normal' life should be and what the West can do to rescue the rest of the world from the wrath of nature. In Fritz's (1961, p. 655) terms: 'in its most general sociological sense, a disaster is defined as a basic disruption of the social context within which individuals and groups function, or a radical departure from the pattern of normal expectations'.

The continuing quest for pantometry that we discussed in Chapter 4, however, has led to the increasing importance of quantitative thresholds in defining disasters. The importance of statistics and numbers of all sorts is mirrored in the growing influence of databases that record disasters. These databases not only support the governmentality of disaster (Chapter 5) but also justify many scientific inquiries. To be fully convinced, one need only note how many academic publications begin with a reference to the Emergency Events Database (EM-DAT) or one of the multiple rankings of regions and countries in terms of losses to justify that the location of their study is relevant because it is often affected by disasters or it suffers significant losses (or both).

As such, the invention of disaster becomes statistical and disaster studies a branch of numerology where statistical thresholds in the impact of natural hazards define whether the occurrence of a particular natural phenomenon turns into a disaster, whether the impact falls beyond the 'normal', and whether it deserves external attention and intervention. Databases and statistical thresholds thus become the basis for drawing a line between the normal and a disaster where the normal is set by the average impact of natural hazards in the West and the expectations of the project of modernity. In this process, those who are affected have virtually no say although the significance of the impact is inherently relative and subjective.

The invention of disaster is also inherently spatial and social. Indeed, the thresholds set to delineate disaster and separate them from acceptable losses, the *dis*-aster, apply to areas and social units, at different scales, so that the labelling of a particular event or process as a disaster become sweeping, although personal/family suffering is much easier to pinpoint than collective grief. As Canguilhem (1966, p. 156, our brackets) adds:

La frontière entre le normal et le pathologique [i.e. a disaster] est imprécise pour des individus multiples considérés simultanément, mais elle est parfaitement précise pour un seul et même individu considéré successivement. Ce qui est normal, pour être normatif dans des conditions données, peut devenir pathologique [i.e. a disaster] dans une autre situation, s'il se maintient identique à soi. De cette transformation c'est l'individu qui est juge parce que c'est lui qui en pâtit, au moment même où il se sent inférieur aux tâches que la situation nouvelle lui propose[4].

In this perspective, the *dispositif* of disaster risk reduction manufactures and labels 'events' and 'processes' as disasters (or not) based on artificial criteria that reflect the expectations of the project of modernity and its normative injunctions. Events and processes therefore are promoted to, and demoted from, the status of disaster on the basis of the number of people killed, the number of individuals affected, and the economic value of losses. In this regard, it is similar to the way in which the International Classification of Diseases (ICD) of the World Health Organization creates pathologies by setting thresholds in physiological processes or in which the Diagnostic and Statistical Manual (DSM) of the American Psychiatric Association arbitrarily establishes that some forms of behaviours are mental disorders and others are not. Indeed, the very evidence that the contents of both ICD and DSM are constantly changing shows that there is no irrefutable fact or reality behind any of the diseases and mental disorders listed in these reports. As Canguilhem (1966) brilliantly exposed, the construction of *dis*-eases, mental *dis*-orders and *dis*-asters as *dis*-functions of the body, the mind and society all respond to the same injunctions associated with the normative nature of everyday life as imposed by Western values, which are those of the project of modernity and more particularly neoliberalism (Chapter 5). In fact, one can clearly spot such framing of *dis*-aster at the onset of the Age of Enlightenment, more particularly in Defoe's (1704, p. 55) collection of accounts of the storm that hit England and Wales in 1703:

> Some Gentlemen, whose Accounts are but of common and trivial Damages, we hope will not take it ill from the Author, if they are not inserted at large; for that we are willing to put in nothing here common with other Accidents of like nature; or which may not be worthy of a History and a Historian to record them; nothing but, what may serve to assist in convincing Posterity that this was the most violent Tempest the World ever saw.

The contemporary statistical evidence for the existence of disasters and their growing or decreasing occurrence, as for diseases and mental disorders, is usually presented in tables, graphs and maps of all sorts which are designed to provide irrefutable and simple proof to scholars, policy-makers and society as a whole that disasters are a thing and that they are an impediment to the project of modernity. In fact, tables, graphs and maps are meant to capture spatial heterogeneities and temporal trajectories. As Canguilhem (1966, p. 114, our brackets again) states, '*l'anomalie éclate dans la multiplicité spatiale*', which is that the West sets the norm because (its) numbers show that it is less affected, while '*la maladie* [i.e. disasters] *éclate dans la succession chronologique*'[5], which is that (1) disasters mark an interruption in the process of 'development' as envisioned by the project of modernity and (2) disasters have occurred more frequently over the past century and hence deserve more scientific and policy attention.

However, delineation and the very existence of such statistical thresholds, expressed visually through tables, graphs and maps, do not reflect any objective reality or 'transcendental signifier', in Derrida's lexis. Rather, it mirrors a normative perspective of what life and the world should be: a view from the West inherited from the Enlightenment and the project of modernity, which is that nature should not be a threat nor have any harmful impact on people. It is the *sine qua non* condition for the prosperity of societies. We find a brilliant analogy of the importance of such normative approach to life in Canguilhem (1966, pp. 112–113, our brackets again):

> Il y a une polarité dynamique de la vie. Pour autant que les variations morphologiques ou fonctionnelles [i.e. the occurrence of natural hazards] sur le type spécifique [i.e. people's everyday life] ne contrarient pas ou n'invertissent pas cette polarité, l'anomalie est un fait toléré; dans le cas contraire, l'anomalie est ressentie comme ayant valeur vitale négative et elle se traduit extérieurement comme telle. C'est parce qu'il y a des anomalies vécues ou manifestées comme un mal organique [i.e. a disaster] qu'il existe un intérêt affectif d'abord, théorique ensuite, pour les anomalies. C'est parce que l'anomalie est devenue pathologique [i.e. a disaster] qu'elle suscite l'étude scientifique des anomalies. De son point de vue objectif, le savant ne veut voir dans l'anomalie que l'écart statistique, en méconnaissant que l'intérêt scientifique du biologiste [i.e. any scholar of disaster] a été suscité par l'écart normatif. En bref, toute anomalie n'est pas pathologique [i.e. disaster], mais seule l'existence d'anomalies pathologiques [i.e. disasters] a suscité une science spéciale des anomalies [i.e. disaster studies] qui tend normalement, du fait qu'elle est science, à bannir de la définition de l'anomalie toute implication de notion normative. Les écarts statistiques qui sont les simples variétés ne sont pas ce à quoi on pense quand on parle d'anomalies, mais les difformités nuisibles ou même incompatibles avec la vie sont ce à quoi on pense, en se référant à la forme vivante ou au comportement du vivant non pas comme à un fait statistique, mais comme à un type normatif de vie[6].

From this perspective, we agree with Hewitt (1983, p. 14) that 'it [i.e. hazard research also known nowadays as disaster studies] has invented its problem field to suit its convenience'. This convenience provides a goal for the project of modernity as framed in the eighteenth century and revived by the emergence of climate change in the public arena (Chapter 6). The quest for consensual and universal definitions of key concepts such as disaster, hazard, vulnerability, capacities and resilience (Chapter 1) thus seems like an anxious attempt at claiming and legitimising a field of research and *dispositif* of governmentality that do not make much sense outside of the West. The litany of glossaries and definitions available in the field appears as a desperate attempt at rationalising, categorising, structuring and normalising what cannot be, because, again, there is no referent and no single

transcendental reality beyond the concept. Worse, the hegemonic logocen-
trism of the West has prevented other understandings and representations
of the world to flourish (Chapter 2).

On pluralism and the *petits récits* of disaster

In fact, these other understandings of the world have been suppressed
because they do not meet the expectations of Western science in that they
do not fit the normative prospects of the project of modernity and the
life it entails. These other world views/senses have been neglected, repudi-
ated or romanticised as per the discourse on local/indigenous/traditional
knowledge (Chapter 6). In Lyotard's (1979, p. 48) terms:

> Le scientifique s'interroge sur la validité des énoncés narratifs, et con-
> state qu'ils ne sont jamais soumis à l'argumentation et à la preuve. Il
> les classe dans une autre mentalité: sauvage, primitive, sous-développée,
> arriérée, aliénée, faite d'opinions, de coutumes, d'autorité, de préjuges,
> d'ignorances, d'idéologies. Les récits sont des fables, des mythes, des
> légendes, bons pour les femmes et les enfants. Dans les meilleurs cas, on
> essaiera de faire pénétrer la lumière dans cet obscurantisme, de civiliser,
> d'éduquer, de développer. Cette relation inégale est un effet intrinsèque
> des règles propres à chaque jeu. On en connait les symptômes. C'est
> toute l'histoire de l'impérialisme culturel depuis les débuts de l'Occi-
> dent. Il est important d'en reconnaitre la teneur qui le distingue de tous
> les autres: il est commandé par l'exigence de légitimation[7].

Considering disaster a Western invention, with no single and universal
transcendental reality, legitimised by fabricated 'facts' that suit the purpose of the project of modernity, confronts the very foundations of disaster studies inherited from the Enlightenment and firmly grounded in
the modern *épistémè* of the West. It challenges the very nature of reason
and its ability to inform understanding of what we call disaster. It under-
mines the ontological and epistemological assumptions that underpin the
predominantly positivist and reductionist approach to apprehending and
addressing disaster (Chapter 4). It ultimately contests the relevance of the
project of modernity and its goal to free people from the danger of nature.
Indeed, as Quijano (1992, p. 20) once asked:

> Pues nada menos racional, finalmente, que la pretensión de que la
> específica cosmovisión de una etnia particular sea impuesta como la
> racionalidad universal, aunque tal etnia se llame Europa Occidental.
> Porque eso, en verdad, es pretender para un provincianismo el título
> de universalidad[8].

It is in this context that '*la crítica del paradigma europeo de la racionalidad/
modernidad es indispensable, más aún, urgente*'[9] (Quijano, 1992, p. 19). This is

a call for the end of any *métarécit* in disaster studies or, as Foucault (1984a, p. 46) once suggested, 'this means that the historical ontology of ourselves must turn away from all projects that claim to be global or radical'.

The end of any *métarécit* on disaster entails the emergence of multiple *petits récits*, which are to reflect the plurality of cultures and world views/ senses, the plurality of ways of knowing, and epistemologies in understanding what we usually call disaster (should there be such a thing as a disaster in the first place). Could there thus be any other interpretations outside of the West? Could multiple understandings co-exist?

Indeed, our intention is not to challenge the rationale behind and the relevance of the Western discourses on disaster within their cradle, which is the West. Nor is it our intention to dismiss people's suffering or the impact of natural phenomena on people's everyday lives. Our point is about challenging a unique, allegedly universal and monolithic interpretation of such realities and their mediation through Western concepts such as disaster which, we argue, do not reflect everyone's understanding of the world across all cultures.

Indeed, we contend that different cultures may hold different interpretations and these interpretations are mediated by local world views/senses and ultimately are translated into concepts through words specific to local languages if we are to assume, in the first place, that concepts and languages are relevant anchors for such a process of mediation. 'This does not mean a rejection of Western categories but signals the beginning of a new and autonomous relation to them' (Das, 1989, p. 310). A co-existence of multiple *épistémès* that do not need to seek legitimisation from each other.

We therefore disagree with Levinas (1972, p. 79) that this co-existence of *épistémès* would be chaotic, that '*la sarabande des cultures innombrables et équivalentes, chacune se justifiant dans son propre contexte, crée un monde, certes, dés-occidentalisé, mais aussi un monde désorienté*'[10]. In fact, it is our contention that 'fears of relativism are prompted more by perceived dangers to academic turfs than any "real" relativist threat' (Agrawal, 1995, p. 427). We see the co-existence of multiple *épistémès* as a possible future if it draws on the recognition of diversity and otherness, on respect and trust, and on a genuine postcolonial agenda that is neither rigid in time and space nor set in the stone of any form of definite and unique reason. As Mbembe (1992, p. 8) notes: 'the postcolony is chaotically pluralistic and that it is in practice impossible to create a single, permanently stable system out of all the signs, images and markers current in the postcolony'.

One may see this agenda as an opportunity for liberation. Not a liberation from the threat of nature but a liberation from the oppression of Western ontologies and epistemologies that have prevented other world views/senses and ways of knowing to emerge and flourish. A liberation of *subjugated knowledges* to use Foucault's expression. Or, in Quijano's (1992, p. 19) terms:

> Lo que hay que hacer es algo muy distinto: liberar la producción del conocimiento, de la reflexión y de la comunicación, de los baches de la racionalidad/modernidad europea. Fuera de "Occidente", en

virtualmente todas las culturas conocidas, toda cosmovisión, todo imaginario, toda producción sistemática de conocimiento, están asociadas a una perspectiva de totalidad. Pero en esas culturas, la perspectiva de totalidad en el conocimiento, incluye el reconocimiento de la heterogeneidad de toda realidad; de su irreductible carácter contradictorio; de la legitimidad, esto es, la deseabilidad, del carácter diverso de los componentes de toda realidad, y de la social en consecuencia. Por lo tanto, la idea de totalidad social, en particular, no solamente no niega, sino que se apoya en la diversidad y en la heterogeneidad histórica de la sociedad, de toda sociedad. En otros términos, no solamente no niega, sino requiere la idea del "otro", diverso, diferente[11].

Re/constructing disaster and postcolonial disaster studies

Horkheimer (1947, pp. 124–125) once argued that 'the disease of reason is that reason was born from man's urge to dominate nature, and the "recovery" depends on insight into the nature of the original disease'. We hope that this book has shed light on the 'original disease' of disaster studies and that it has contributed to an exercise of deconstruction, which is to disassemble disaster as a concept, disaster studies as a field of scholarship, and disaster risk reduction as a *dispositif* of governmentality, all firmly grounded in Western science and the Western art of government.

It is time to re/construct our field of scholarship from a postcolonial and pluralistic perspective and to make the *petits récits* of what we call disaster emerge. It may seem a purely academic exercise but it is more. To paraphrase Lenin: there cannot be any significant change and more grounded approaches in how we deal with what the West calls disaster until there are solid alternative theories, until other forms of understandings emerge and until the everyday forms of *contre conduites* and resistance that we discussed in Chapters 8 and 9 are recognised and considered. As such, re/construction is not about re/building from scratch.

In fact, a re/construction agenda is not necessarily about completely abandoning the Western invention, '*an impossible aim*' according to Clifford (1986b, p. 119). It is, however, about 'opening ourselves to different histories' (Clifford, 1986b, p. 119). Re/construction, as a follow-up to deconstruction, is thus about fostering multiple 'inventions of disaster' over the hegemony of one single Western set of ontologies and epistemologies. In Said's (1994, pp. 51–52) words: 'For each locale in which the engagement occurs, and the imperialist model is disassembled, its incorporative, universalizing, and totalizing codes rendered ineffective and inapplicable, a particular type of research and knowledge begins to build up'. As such, '*la déconstruction est inventive ou elle n'est pas; elle ne se contente pas de procédures méthodiques, elle fraye un passage, elle marche et marque*'[12] (Derrida, 1987).

We recognise that 'decolonising' disaster studies, as a dialectical agenda set to push back against the current state of scholarship, may be a worthwhile first step, especially in places that are still subjugated to colonial power or bound to neo-/post-colonial ties. In those places, decolonising may serve a purifying and liberating purpose, and one cannot remain insensitive to Tuhiwai-Smith's (2012, p. 24) wholehearted call:

> The reach of imperialism into "our heads" challenges those who belong to colonized communities to understand how this occurred, partly because we perceive a need to decolonize our minds, to recover ourselves, to claim a space in which to develop a sense of authentic humanity.

In fact, it may be required for all of us to first 'unlearn (...) the dominative mode' (Williams, 1960, pp. 355–356) to eventually move on towards a postcolonial agenda.

In the longer run, the postcolonial agenda we propose here is nonetheless about suppressing any dialectical relationship that allows one interpretation of what we call disaster to impose itself over another. Emancipation, as Fanon (1952) once argued, is about breaking away from the dialectical logic of colonialism because, as Memmi (1957, p. 141) reiterated a few years later, '*la condition coloniale ne peut être changée que par la suppression de la relation coloniale*'[13]. It is in this perspective that the endeavour of deconstruction that we have proposed in this book has consistently challenged binaries and dialectical relationships that reinforce a Western interpretation of the world with a view to open up ontological and epistemological opportunities beyond such polarities. To cite Bhabha (1994, p. 246):

> to reconstitute the discourse of cultural difference demands not simply a change of cultural contents and symbols; a replacement within the same time-frame of representation is never adequate. It requires a radical revision of the social temporality in which emergent histories may be written, the rearticulation of the "sign" in which cultural identities may be inscribed.

Our intention is therefore to learn from below (Spivak, 2013), in fact from multiple belows, rather than to solely push back against a single above. As Hewitt (1994, p. 8) has long argued in the specific context of disaster studies:

> among other things, such work seems to involve and require a different modus operandi, methodologies and perspectives: a view from within rather than outside communities, a participation in the sense of crisis. One requires insight rather than oversight; a capacity to listen to, comprehend and interpret experience and circumstances expressed in vernacular language rather than technical ways. In sum, one will have to recognise, assess and express the view from below.

This agenda is likely to be challenging if it is to be envisioned from the West, if the emergence of diverse and pluralistic ontologies and epistemologies is to be considered through the sole perspective of what we already know. Mbembe (2010, p. 182) suggested:

> À distance de la suffisance positiviste, il faut donc relire l'histoire de l'Occident hors d'Occident à rebours du discours occidental sur sa propre genèse, à rebours de ses fictions, de ses évidences parfois vides de contenu, ses déguisements, ses ruses et – cela vaut la peine de le répéter – sa volonté de puissance (qui, comme on vient de le suggérer, est profondément encastrée dans une structure d'impuissance et d'ignorance)[14].

As JanMohamed (1985, p. 65) further argued: 'genuine and thorough comprehension of Otherness is possible only if the self can somehow negate or at least severely bracket the values, assumptions, and ideology of his culture'. Therefore, 'our understanding of disaster needs to be turned inside out and not the other way around, as it tends to become, thanks to the "expert" notions of what is a disaster' (Jigyasu, 2005, p. 59).

Genuine participatory pluralism probably offers the most promising opportunities among the array of approaches we reviewed in Chapter 4. It has the potential for opening up spaces for ontological and epistemological alternatives in researching disaster, whether it is actually a relevant issue and whether there is such a concept as there is in the West. However, supporting the ethos that underpins participatory pluralism entails, by nature, that there is no single methodological pathway, nor any standard toolkit, including those usually associated with Participatory Learning and Action, to begin with. Methods for researching what we usually mean by disaster require us to reflect local world views/senses, that is to think, be and act by oneself and for oneself (Senghor, 1971; Eboussi-Boulaga, 1977; Salazar, 1991). They will be multiple, possibly unique and hence hardly comparable or transferable.

In this perspective, as for any postcolonial agenda (Fanon, 1952; Césaire, 1956; Memmi, 1957; Eboussi-Boulaga, 1977; Spivak, 1993b; Diagne, 2013), it is again essential to recognise the importance of language because any interpretation of what we mean by disaster and any discourse on how people deal with natural phenomena such as earthquakes, cyclones and volcanic eruptions are likely to be mediated by words and concepts. Our alternative understandings of disaster therefore should start with local language, whether it is written or not. In Gramsci's (1929–35, Q10, XXXIII, §44) terms:

> Posta la filosofia come concezione del mondo e l'operosità filosofica non concepita più [solamente] come elaborazione "individuale" di concetti sistematicamente coerenti ma inoltre e specialmente come lotta culturale per trasformare la "mentalità" popolare e diffondere le innovazioni filosofiche che si dimostreranno "storicamente vere" nella misura in cui diventeranno concretamente cioè storicamente e socialmente universali, la questione del linguaggio e delle lingue "tecnicamente" deve essere posta in primo piano[15].

Genuine participatory pluralism intrinsically poses challenges to those who are interested in comparing studies to draw theoretical lessons, in scaling up insights learnt in a very particular locality to inform international policy, or in simply fostering cross-cultural learning among people confronted with the same issues. The mediation of learning across culture, especially between the West and the rest of the world, has received considerable attention in anthropological literature since the early charges against Western ethnography led by Asad (1973), Rabinow (1977, 1996), and Clifford and Marcus (Clifford and Marcus, 1986; Marcus and Fischer, 1986; Clifford, 1988) in the 1970s and 1980s. Marcus and Fischer (1986, p. 31) then wrote:

> In cross-cultural communication, and in writing about one culture for members of another, experience-near or local concepts of the cultural other are juxtaposed with the more comfortable, experience-far concepts that the writer shares with his readership. The act of translation involved in any act of cross-cultural interpretations is thus a relative matter with an ethnographer as mediator between distinct sets of categories and cultural conceptions that interact in different ways at different points of the ethnographic process.

Anthropologists have actually shown us that overcoming these challenges is often a matter of fair dialogue between all parties involved in any endeavours of research (Marcus and Fischer, 1986; Clifford, 1988).

A call for dialogue and alliance

Indeed, fostering a postcolonial agenda into disaster studies calls for a reconsideration of the power relations that currently underpin our field of scholarship (Chapters 2 and 3). It inherently puts local researchers, including local people/survivors whose agency as potential researchers has to be recognised, and/or those researchers grounded in the particular place they study, in the front seat. As Hewitt (1995b, p. 330) once suggested: 'letting those in hazard speak for and of themselves, is one of the few possibilities for keeping the faces and pain in the foreground of interpretation and response'.

Although they are facing many challenges of their own, which we fully acknowledge and do not mean to underestimate (e.g. Padilla-Goodman, 2010; Miyazawa, 2018; Balay-as, 2019; Takakura, 2019), these researchers are still likely to be those who have the best understanding of local societies, local world views/senses, and local ways of knowing, all of which are essential dimensions of a postcolonial agenda. They are probably those who will have enough knowledge of local languages to most appropriately reflect local world views/senses and hence better translate the realities of everyday life. Also, local and grounded researchers will likely be those who know better than anyone else what issues need to be studied, where the priorities should be and whether these centre on what we call disaster.

It is not our intention, however, to contribute here to the debate on positionality and legitimacy in the postcolonial space. Postcolonial literature has covered this issue extensively, especially since Spivak's seminal piece on whether the subaltern can speak (e.g. Spivak, 1987, 1988, 1990b; Bhabha, 1994; Chakrabarty, 2000). It has helped move the debate beyond Hegel's (1807) dialectic of independence and dependence of self-consciousness or beyond a polarisation of relationships between insiders and outsiders. Self-consciousness indeed transcends localities, national borders and passport affiliations (Chakrabarty, 2000), and there is no single, easy answer to who should legitimately conduct research in any specific place. In one of his most famous quotes, Gramsci (1929–35, Q11, XVIII, §12) reminded us that

> l'inizio dell'elaborazione critica è la coscienza di quello che è realmente, cioè un "conosci te stesso" come prodotto del processo storico finora svoltosi che ha lasciato in te stesso un'infinità di tracce accolte senza beneficio d'inventario. Occorre fare inizialmente un tale inventario[16].

As such, our intention is not to foster an exclusive approach similar to the one that characterises the practice of inclusion in disaster risk reduction, which we critiqued in Chapter 6. Indeed, it is not our intention to replace one hegemonic *récit* by another, including at the very local level, and one hegemonic way of approaching scholarship by another. Parry (1987, p. 28) nicely summarised that

> a reverse discourse replicating and therefore reinstalling the linguistic polarities devised by a dominant centre to exclude and act against the categorized, does not liberate the 'other' from a colonized condition where heterogeneity is repressed in the monolithic figures and stereotypes of colonialist representation, and into a free state of polymorphous native 'difference'.

Rather, it is our intention to encourage a fair dialogue among researchers. A dialogue that encourages local or grounded researchers (or both), in cases where local voices are suppressed or unheard for any reason, including totalitarian regimes and other forms of repression, to take the lead. Local or grounded researchers (or both) are to define the research agenda, to provide the most appropriate ontological and epistemological framing, and eventually to carry out and valorise the research. This is obviously assuming that these canons of research are locally relevant, which may not be the case.

Fostering local and/or grounded research leadership does not preclude outside researchers, in all their diversity, from contributing. In the contemporary research landscape, they are those who may be able to access required resources and equipment, to leverage power relations with and among the stakeholders of what we call disaster risk reduction so that the latter progressively buy in to whatever changes postcolonial scholarship may propose.

Moving our agenda forward, therefore, entails playing within the current system, cracking into its interstices and maximising opportunities. It is about elevating the everyday forms of *contre conduites* and resistance (Chapters 8 and 9) that we observe on the ground onto our field of scholarship. It is a tactical game to subvert the neoliberal system of contemporary research. According to Tuhiwai-Smith (2012, p. 199),

> [s]truggle can be mobilized as resistance and as transformation. It can provide the means for working things out 'on the ground', for identifying and solving problems of practice, for identifying strengths and weaknesses, for refining tactics and uncovering deeper challenges.

The roots of such an agenda may be found in Gramsci's tactical approach (his so-called '*guerra di posizione*' or 'war of position') to topple a hegemonic form of domination, which, in his own experience, was both social/economic (between proletariat and bourgeoisie) and geographical (between southern and northern Italy). Although we reject the military language and recognise that Gramsci's approach is largely (but not exclusively) dialectical in that it aims at bringing the proletariat to power or southern Italy to challenge the economic and political power of the north, it provides a very useful framework to guide our agenda. It is, indeed, an approach that is based on alliances and dialogue and resonates with many postcolonial scholars (e.g. Chakrabarty, 1995; Young, 2004).

Gramsci (1930) recognised that multiple different perspectives may co-exist and have common interests outside of the dominant hegemony. In the 1920s, the views of the urban proletariat of northern Italy differed from those of the peasants of the south. Yet they had a common goal to challenge the hegemony of the bourgeoisie. It is through the alliance of these different groups, Gramsci argued, that such a goal was achievable. Essential to our postcolonial agenda is that this alliance accommodates cultural and linguistic heterogeneity, so that it goes beyond a simply dialectical approach.

Forging such an alliance requires a fair dialogue and solidarity so that all parties get to recognise and accept shared aspirations as well as cultural differences. In Freire's (1963, p. 12) terms: '*Somente um método dialogal, ativo, participante, poderia realmente fazê-lo* [i.e. foster an active education]. *Somente pelo diálogo que, nascendo numa matriz crítica, gera criticidade e que implica numa relação de como conseguir esses objetivos*'[17]. The dialogue Freire (1968) eventually developed draws on love (i.e. compassion for the others and the world that surrounds us), humility, faith (i.e. trust that the others can lead), hope and critical thinking. Chakrabarty (1995, p. 756) adds:

> to be open-ended, I would argue, a dialogue must be genuinely non-teleological; that is, one must not presume, on any a priori basis,

that whatever position our political philosophy/ideology suggests as correct will be necessarily vindicated as a result of this dialogue. For a dialogue can be genuinely open only under one condition: that no party puts itself in a position where it can unilaterally decide the final outcomes of the conversation

This alliance is to build upon consciousness, among researchers, that change is possible when it comes to studying what we call disasters. Consciousness that disasters do not necessarily sit within the nature/hazard–culture/vulnerability binary. Consciousness that Western ontologies and epistemologies inherited from the Enlightenment are not the only way forward and that English and other European languages are not the only languages that convey reason and truth.

Raising wide consciousness among researchers of all backgrounds, of all cultures, and of all regions of the world is absolutely critical. As Gramsci (1930) contended, '*nessuna azione di massa è possibile se la massa stessa non è convinta dei fini che vuole raggiungere e dei metodi da applicare*'[18]. Freire (1963, 1968, 1970) famously referred to this process of raising consciousness among those who have been oppressed as *conscientização*, which, in Freire's own terms (1970, p. 452), 'refers to the process in which men, not as recipients, but as knowing subjects, achieve a deepening awareness both of the socio-cultural reality which shapes their lives and of their capacity to transform that reality'. For Freire, this process is grounded in people's (in our case, researchers') own experience of the world and, in the specific context of disaster studies, of the current, skewed research landscape that favours some researchers to the detriment of others. It is indeed only through a critical reflection and realisation of the root causes of unequal power relations, Freire (1968, 1970) argued, that those who are oppressed will endeavour to act and contribute to their liberation from such oppression. We strongly believe that there are enough brilliant scholars in Asia and the Pacific, Africa and Latin America to lead this process and raise consciousness among their peers.

Utopia or foreseeable future?

Our agenda may seem utopic. Many may believe so when contemplating the contemporary research landscape that we briefly discussed in Chapter 2. Many may think that the issues we expose in this book are so entrenched that any attempts at challenging them would be in vain. Some may think that reversing, or at least balancing, power and knowledge relations in studying disaster is a chimera. We do not believe so.

Our hopes lie in the emerging momentum that recent publications and projects have generated. They lie in the level of critical consciousness that many, especially younger, scholars have recently expressed by engaging with critical theory when researching what we call disaster and in the discussions that occurred during recent meetings, especially, once again, with the up-and-coming generation of scholars. They lie in the traction that the manifesto *Power, Prestige and Forgotten Values* in disaster studies has garnered and again in the fact that open-ended projects like the *Gender Responsive Resilience & Intersectionality in Policy and Practice* initiative are being funded by large research schemes. Our hopes lie in the fact that many of these creative projects build upon ethical agreements that foster fair and respectful rapports between scholars and in that publishers are opening up opportunities for articles, books and other forms of media that challenge the standards of Western scholarship. These are all signs of hope that one cannot easily dismiss. Signs that many of us are ready to '*experience the impossible*' (Derrida, 1987) and to cross over the long-established and rigid boundaries of disaster studies as we know it today. Signs that many of us agree with Delphy (1991, p. 89, 96) that

> le courage d'affronter l'inconnu est la condition de l'imagination et que la capacité d'imaginer un monde autre est un élément essentiel de la démarche scientifique' and that 'pour avancer, il faut d'abord renoncer à certaines évidences; ces "évidences" procurent le sentiment confortable que procurent toutes les certitudes mais elles nous empêchent de poser des questions, ce qui est sinon la seule, au moins la plus sûre façon de parvenir à des réponses[19].

In addition, there are examples to build upon and these examples are not rare nor isolated. They span a range of fields and disciplines and, as such, constitute powerful inspirations for a field of scholarship such as disaster studies that claims to be multidisciplinary (see Connell, 2007; de Sousa Santos, 2007 for broad and theoretical reviews). These include, for example, the broad spaces of African philosophy (e.g. Hountondji, 1976; Eboussi-Boulaga, 1977; Wiredu, 1980; Mudimbe, 1988), Latin American psychology (e.g. Martin-Baró, 1994) and Pacific studies (e.g. Vaioleti, 2006). There are also multiple local examples such as *Sikolohiyang Pilipino* (e.g. Pe-Pua, 1982; Enriquez, 1992) and *Bagong Kasaysayan* in the Philippines (e.g. Salazar, 1991; Guillermo, 2009), *kaupapa Māori* in Aotearoa (e.g. Tuhiwai-Smith, 2012; Hoskins and Jones, 2017), Indian psychology (e.g. Safaya, 1976) and, of course, South Asian history (e.g. Guha, 1997; Chakrabarty, 2000). Yet it is possibly Oyěwùmí's (1997) critique and re/conceptualisation of what the West calls gender in the context of her own Yoruba society that constitute one of the most powerful examples. It is the archetype of the kind of postcolonial studies that should guide our future scholarship of disaster should one dare to take our agenda forward.

Going further down this road is going to be a long and challenging journey but one that is essential and that will be immensely rewarding. We believe that our agenda and this book in particular are going to stir scepticism and criticism. We hope that critiques will be fair, respectful and constructive, for the sake of moving our field forward rather than digging in one's heels. On the other hand, those of us who are willing to take on this agenda should be ready to accept these critiques and face opposition. A constructive dialogue is the only way forward.

To conclude, let us all remember what Spivak (1990b, p. 4) once said: 'if we want to start something, we must ignore that our starting point is shaky. If we want to get something done, we must ignore that the end will be inconclusive'.

Notes

1 *'the signified concept is never present in and of itself, in a sufficient presence that would refer only to itself. Essentially and lawfully, every concept is inscribed in a chain or in a system within which it refers to the other, to other concepts, by means of the systematic play of differences. Such a play, différance, is thus no longer simply a concept, but rather the possibility of conceptuality, of a conceptual process and system in general'.* From the English edition translated by Alan Bass and published by University of Chicago Press in 1982.

2 *'There is nothing outside of the text [there is no outside-text]'.* From the English edition translated by Gayatri Chakravorty Spivak and published by the Johns Hopkins University Press in 1974.

3 *'Nobody today thinks that the word "dis-aster" is connected with astrology or can claim to be misled about the opinions of someone who uses the word'.* From the English edition translated by Quintin Hoare and Geoffrey Nowell Smith and published by International Publishers in 1971.

4 *'The borderline between the normal and the pathological is imprecise for several individuals considered simultaneously but it is perfectly precise for one and the same individual considered successively. In order to be normative in given conditions, what is normal can become pathological in another situation if it continues identical to itself. It is the individual who is the judge of this transformation because it is he who suffers from it from the very moment he feels inferior to the tasks which the new situation imposes on him'.* From the English edition translated by Carolyn R. Fawcett and published by Zone Books in 1991.

5 *'An anomaly manifests itself in spatial multiplicity, disease, in chronological succession'.* From the English edition translated by Carolyn R. Fawcett and published by Zone Books in 1991.

6 *'There is a dynamic polarity of life. As long as the morphological or functional variations on the specific type do not hinder or subvert this polarity, the anomaly is a tolerated fact; in the opposite case the anomaly is felt as having negative vital value and is expressed as such on the outside. Because there are anomalies which are experienced or revealed as an organic disease, there exists first an affective and then a theoretical interest in them. It is because the anomaly has become pathological that it stimulates scientific study. The scientist, from his objective point of view, wants to see the anomaly as a mere statistical divergence, ignoring the fact that the biologist's scientific interest was stimulated by the normative divergence. In short, not all anomalies are pathological but only the existence of pathological anomalies has given rise to a special science of anomalies which, because it is science, normally tends to rid the definition of anomaly of every implication of a normative idea. Statistical divergences such as simple varieties are not what one thinks of when one speaks of anomalies; instead one thinks of harmful deformities or those even incompatible with life, as one refers to the living form or behavior of the living being not as a statistical fact but as a normative type of life'.* From the English edition translated by Carolyn R. Fawcett and published by Zone Books in 1991.

7 *'The scientist questions the validity of narrative statements and concludes that they are never subject to argumentation or proof. He classifies them as belonging to a different mentality: savage, primitive, underdeveloped, backward, alienated, composed of opinions, customs, authority, prejudice, ignorance, ideology. Narratives are fables, myths, legends, fit only for women and children. At best, attempts are made to throw some rays*

of light into this obscurantism, to civilize, educate, develop. This unequal relationship is an intrinsic effect of the rules specific to each game. We all know its symptoms. It is the entire history of cultural imperialism from the dawn of Western civilization. It is important to recognize its special tenor, which sets it apart from all other forms of imperialism: it is governed by the demand for legitimation'. From the English edition translated by Geoff Bennington and Brian Massumi and published by University of Minnesota Press in 1984.

8 *'Nothing is less rational, finally, than the pretension that the specific cosmic vision of a particular ethnie should be taken as universal rationality, even if such an ethnie is called Western Europe because this is actually pretend to impose a provincialism as universalism'.* From the English edition published in Cultural Studies in 2007.

9 *'The critique of the European paradigm of rationality/modernity is indispensable even more, urgent'.* From the English edition published in Cultural Studies in 2007.

10 *'the world created by this saraband of countless equivalent cultures, each one justifying itself in its own context, is certainly dis-Occidentalized; however, it is also disoriented'.* From the English edition translated by Nidra Poller and published by University of Illinois Press in 2006.

11 *'What is to be done is something very different: to liberate the production of knowledge, reflection, and communication from the pitfalls of European rationality/modernity. Outside the 'West', virtually in all known cultures, every cosmic vision, every image, all systematic production of knowledge is associated with a perspective of totality. But in those cultures, the perspective of totality in knowledge includes the acknowledgement of the heterogeneity of all reality; of the irreducible, contradictory character of the latter; of the legitimacy, i.e., the desirability, of the diverse character of the components of all reality and therefore, of the social. The idea of social totality, then, not only does not deny, but depends on the historical diversity and heterogeneity of society, of every society. In other words, it not only does not deny, but it requires the idea of an 'other' – diverse, different'.* From the English edition published in Cultural Studies in 2007.

12 *'Deconstruction is inventive or it is nothing at all; it does not settle for methodical procedures, it opens up a passageway, it marches ahead and leaves a trail'.* From the English edition translated by Peggy Kamuf and Elizabeth Rottenberg and published by Stanford University Press in 2007.

13 *'The colonial condition cannot be changed except by doing away with the colonial relationship'.* From the English edition translated by Howard Greenfeld and published by Plunkett Lake Press in 2013.

14 *'Taking distance from positivist conceits, we must therefore reread the history of the West against the grain of Western accounts of its own genesis, reading against its fictions, its obvious and sometimes empty truths, its disguises, its ploys, and – it bears repeating – its will for power'.* From the English translation published in Public Culture in 2011.

15 *'We have established that philosophy is a conception of the world and that philosophical activity is not to be conceived solely as the "individual" elaboration of systematically coherent concepts, but also and above all as a cultural battle to transform the popular "mentality" and to diffuse the philosophical innovations which will demonstrate themselves to be "historically true" to the extent that they become concretely - i.e. historically and socially – universal. Given all this, the question of language in general and of languages in the technical sense must be put in the forefront of our enquiry'.* From the English edition translated by Quintin Hoare and Geoffrey Nowell Smith and published by International Publishers in 1971.

16 *'The starting-point of critical elaboration is the consciousness of what one really is, and is 'knowing thyself' as a product of the historical process to date, which has deposited in you an infinity of traces, without leaving an inventory. Therefore it is imperative at the outset to compile such an inventory'.* From the English edition translated by Quintin Hoare and Geoffrey Nowell Smith and published by International Publishers in 1971.

17 '*Only a dialogical, active, participatory method can really do this* [i.e. foster an active education]. *It is only through dialogue, which is critical in origin, that it is possible to generate criticality and facilitate a relationship that allows to achieve these goals*'. Our translation.

18 '*No mass action is possible, if the masses in question are not convinced of the ends they wish to attain and the methods to be applied*'. From the English edition translated by Quintin Hoare and published by Lawrence & Wishart in 1977.

19 '*Having the courage to confront the unknown is a precondition for imagination, and the capacity to imagine another world is an essential element in scientific progress*' and that '*to advance, we must first renounce some truths. These 'truths' make us feel comfortable, as do all certainties, but they stop us asking questions - and asking questions is the surest, if not the only way of getting answers*'. From the English translation published in Women's Studies International Forum in 1993.

11 Postscript

Where to from here?

The agenda put forward in this book is somehow easy to assemble and acknowledge. It is likely to be more challenging to take forward. As Memmi (1957, p. 46) once said, *'refuser la colonisation est une chose, adopter le colonisé et s'en faire adopter en semblent d'autres, qui sont loin d'être liées'*[1]. Similarly, it is one thing to recognise that disaster studies and the governmentality of disaster are skewed towards Eurocentric/Western knowledge and discourses. It is another to actually revisit disaster studies from diverse ontological and epistemological perspectives so that it contributes to reshaping policies and actions towards reducing the risk of disaster across the world. Throughout this book, I have tried to expose some of the challenges inherent to the academic environment we might face when advancing such an agenda. I will not reiterate them here. Nor will I re-emphasise the perspectives and pathways I laid out in the previous chapter. I just hope that many of the 500 signatories of our manifesto *Power, Prestige and Forgotten Values*, which calls for rethinking our research agenda through more respectful, reciprocal and genuine relationships in disasters studies, will dare to 'experience the impossible', to stick to Derrida's (1987) words. It is my own intention to give it a go, in a hopefully not-so-distant future, in a sequel to this book. A sequel that should be the story of mighty crocodiles and more...

Note

1 *'To refuse colonization is one thing; to adopt the colonized and be adopted by them seems to be another; and the two are far from being connected'*. From the English edition translated by Howard Greenfeld and published by Earthscan Publications in 2003.

DOI: 10.4324/9781315752167-11

References

Abinales N. (2012) From disaster response to poverty alleviation: a community initiative to reduce the risk of disasters in the Philippines. In Wisner B., Gaillard J.C., Kelman I. (Eds) *Handbook of hazards and disaster risk reduction.* Abingdon, UK: Routledge, p. 717.

Agozino B. (2003) *Counter-colonial criminology: a critique of imperialist reason.* London, UK: Pluto Press.

Agozino B. (2004) Imperialism, crime and criminology: towards the decolonization of criminology. *Crime, Law and Social Change* 41(4): 343–358.

Agrawal A. (1995) Dismantling the divide between indigenous and scientific knowledge. *Development and Change* 26(3): 413–439.

Agrawal A. (2002) Indigenous knowledge and the politics of classification. *International Social Science Journal* 54(173): 287–297.

Agrawal A., Gibson C.C. (1999) Enchantment and disenchantment: the role of community in natural resource conservation. *World Development* 27(4): 629–649.

Ahmad A. (1992) *In theory: classes, nations, literatures.* London, UK: Verso.

Aitsi-Selmi A., Blanchard K., Al-Khudhairy D., Ammann W., Basabe P., Johnston D., Ogallo L., Onishi T., Renn O., Revi A., Roth C., Peijun S., Schneider J., Wenger D., Murray V. (2015) *United Nations International Strategy for Disaster Reduction Science and Technical Advisory Group report 2015: science is used for disaster risk reduction.* Geneva, Switzerland: United Nations International Strategy for Disaster.

Alburo-Cañete K.Z. (2021) Photo *Kwento*: co-constructing women's narratives of disaster recovery. *Disasters* 45(4): 887–912.

Alexander B. (2012) Hazards and disasters represented in music. In Wisner B., Gaillard J.C., Kelman I. (Eds) *Handbook of hazards and disaster risk reduction.* Abingdon, UK: Routledge, pp. 131–141.

Alexander D. (2002a) *Principles of emergency planning and management.* Oxford, UK: Oxford University Press.

Alexander D. (2002b) From civil defence to civil protection – and back again. *Disaster Prevention and Management* 11(3): 209–213.

Alexander D. (2013) Resilience and disaster risk reduction: an etymological journey. *Natural Hazards and Earth System Sciences* 13: 2707–2716.

Alexander D. (2014) Social media in disaster risk reduction and crisis management. *Science and Engineering Ethics* 20(3): 717–733.

Alexander D., Gaillard J.C., Kelman I., Marincioni F., Penning-Rowsell E., van Niekerk D., Vinnell L. (2020) Academic publishing in disaster risk reduction: past, present, and future. *Disasters* 45(1): 5–18.

Altbach P.G. (2003) *The decline of the guru: the academic profession in developing and middle-income countries.* New York, USA: Palgrave.

Altbach P.G. (2004) Globalisation and the university: myths and realities in an unequal world. *Tertiary Education & Management* 10(1): 3–25.

Althusser L. (1975) *Positions, 1964–1975*. Paris, France: Editions Sociales.

Althusser L. (2018) *Que faire?* Paris, France: Presses Universitaires de France.

Altman D. (1996) Rupture or continuity? The internationalization of gay identities. *Social Text* 48: 77–94.

Amselle J.-L. (2008) *L'Occident décroché: enquête sur les postcolonialismes*. Paris, France: Stock.

Andag R. (2021) *San Julian Pride's Bagyong Yolanda/Super Typhoon Haiyan experience*. Presentation at the *Online Forum on Climate Change-Related Experiences of Non-Normative Genders and Sexualities and their Responses*. 20 January 2021, University of the Philippines Resilience Institute, Online.

Anderson M.B. (1994) Understanding the disaster-development continuum: gender analysis is the essential tool. *Focus on Gender* 2(1):7–10.

Anderson M.B., Woodrow P.J. (1989) *Rising from the ashes: development strategies in times of disasters*. Boulder, USA: Westview Press.

Anderson M.B., Woodrow P.J. (1991) Vulnerability to drought and famine: developmental approaches to relief. *Disasters* 15(1):43–54.

Anderson W.A. (1965) *Some observations on a disaster subculture: the organizational response of Cincinnati, Ohio, to the 1964 flood*. Research note No. 6, Disaster Research Center, The Ohio State University, Columbus, USA.

Anderson-Berry L.J. (2003) Community vulnerability to tropical cyclones: Cairns, 1996–2000. *Natural Hazards* 30(2): 209–232.

Antillano A. (2017) When prisoners make the prison. Self-rule in Venezuelan prisons. *Prison Service Journal* 229: 26–30.

Aragón-Duran E., Lizarralde G., Gonzáles-Camacho G., Oliveira-Ranero A., Bornstein L., Herazo B., Labbé D. (2020) The language of risk and the risk of language: mismatches in risk response in Cuban coastal villages. *International Journal of Disaster Risk Reduction* 50: 101712.

Asad T. (1973) *Anthropology & the colonial encounter*. Amherst, USA: Humanity Books.

Asian Development Bank (2002) *Kiribati: monetization in an atoll society – Managing economic and social change*. Manila, The Philippines: Asian Development Bank.

Asian Development Bank (2009) *Kiribati social and economic report 2008*. Manila, The Philippines: Asian Development Bank.

Bachrach P., Baratz M.S. (1962) Two faces of power. *The American Political Science Review* 56(4): 947–952.

Badan Nasional Penanggulangan Bencana (2010) *National action plan for disaster risk reduction 2010–2012*. Jakarta, Indonesia: Badan Nasional Penanggulangan Bencana.

Baird A., O'Keefe P., Westgate K., Wisner B. (1975) *Towards an explanation and reduction of disaster proneness*. Occasional Paper No 11. Disaster Research Unit, University of Bradford, Bradford, UK.

Bajek R., Matsuda Y., Okada, N. (2008) Japan's jishu-bosai-soshiki community activities: analysis of its role in participatory community disaster risk management. *Natural Hazards* 44(2): 281–329.

Balay-As M. (2019) Disasters through an indigenous lens in the Philippines. PhD thesis, The University of Auckland, Auckland, New Zealand.

Balgos B., Gaillard J.C., Sanz K. (2012) The Warias of Indonesia in disaster risk reduction: the case of the 2010 Mt Merapi eruption. *Gender and Development* 20(2): 337–348.

Ball N. (1975) The myth of the natural disaster. *The Ecologist* 5(10): 368–371.

Bankoff G. (2001) Rendering the world unsafe: 'vulnerability' as a Western discourse. *Disasters* 25(1): 19–35.

Bankoff G. (2003a) Regions of risk: Western discourses on terrorism and the significance of Islam. *Studies in Conflict & Terrorism* 26(6): 413–428.

Bankoff G. (2003b) *Cultures of disaster: society and natural hazard in the Philippines.* London, UK: Routledge.

Bankoff G. (2019) Remaking the world in our own image: vulnerability, resilience and adaptation as historical discourses. *Disasters* 43(2): 221–239.

Banzon Bautista M.C.R. (1993) *In the shadow of the lingering Mt Pinatubo disaster.* Quezon City, The Philippines: University of the Philippines Diliman-College of Social Sciences and Philosophy.

Barahona C., Levy S. (2007) The best of both worlds: producing national statistics using participatory methods. *World Development* 35(2): 326–341.

Barnett J. (2003) Security and climate change. *Global Environmental Change* 13(1): 7–17.

Barrameda T.V., Barrameda A.S.V. (2011) Rebuilding communities and lives: the role of *damayan* and *bayanihan* in disaster resiliency. *Philippine Journal of Social Development* 3: 132–151.

Barrios R (2017) *Governing affect: neoliberalism and disaster reconstruction.* Lincoln, USA: University of Nebraska Press.

Barthes R. (1968) La mort de l'auteur. *Mantéia* 5: 12–17.

Bates T.R. (1975) Gramsci and the theory of hegemony. *Journal of the History of Ideas* 36(2): 351–366.

Beccari B. (2016) A comparative analysis of disaster risk, vulnerability and resilience composite indicators. *PLOS Currents Disasters* 1. 10.1371/currents. dis.453df025e34b682e9737f95070f9b970.

Begum R. (1993) Women in environmental disasters: the 1991 cyclone in Bangladesh. *Gender and Development* 1(1): 34–39.

Bello W. (2001) *The future in the balance: essays on globalization and resistance.* Oakland, USA: Food First Books.

Bello W. (2006) The rise of the relief-and-reconstruction complex. *Journal of International Affairs* 59(2): 281–296.

Benblidia M. (1990) Les priorités de la décennie pour les pays en voie de développement. *UNDRO News* Jan/Feb: 14–17, 21.

Béné C. (2013) *Towards a quantifiable measure of resilience.* Working Paper No. 434, Institute of Development Studies Work, Brighton, UK.

Béné C., Al-Hassan R.M., Amarasinghe O., Fong P., Ocran J. Onumah E., Ratuniata R., Van Tuyen T., Allister McGregor J., Mills D.J. (2016) Is Resilience socially constructed? Empirical evidence from Fiji, Ghana, Sri Lanka, and Vietnam. *Global Environmental Change* 38: 153–170.

Benedicto B. (2008) The haunting of gay Manila: global space-time and the specter of Kabaklaan. *GLQ: A Journal of Lesbian and Gay Studies* 14(2–3): 317–338.

Besnier N. (1994) Polynesian gender liminality through time and space. In Herdt G. (Ed.) *Third sex, third gender: beyond sexual dimorphism in culture and history.* Cambridge, USA: Zone Books, pp. 285–328.

Besnier N., Alexeyeff K. (2014) *Gender on the edge: transgender, gay, and other Pacific islanders.* Honolulu, USA: University of Hawaii Press.

Bhabha H. (1994) *The location of culture.* London, UK: Routledge.

Bhatt M.R. (1998) Can vulnerability be understood? In Twigg J., Bhatt M.R. (Eds) *Understanding vulnerability: South Asian perspectives.* London, UK: Intermediate Technology Publications, pp. 68–77.

Bhatt M.R. (2014) Disaster insurance for the poor. In López-Carresi A., Fordham M., Wisner B., Kelman I., Gaillard J.C. (Eds) *Disaster management: international lessons in risk reduction, response and recovery.* London, UK: Routledge, pp. 178–190.

Bhatt M.R. (2017) Transformation: initiatives towards resilience. *southasiadisasters. net* 160: 2.

Birkmann J. (2006) Indicators and criteria for measuring vulnerability: theoretical bases and requirements. In Birkmann J. (Ed.) *Measuring vulnerability to natural hazards.* Tokyo, Japan: United Nations University Press, pp. 55–77.

Blaikie P., Cannon T., Davis I., Wisner B. (1994) *At risk: natural hazards, people's vulnerability, and disaster.* London, UK: Routledge.

Blaikie P. (2006) Is small really beautiful? Community-based natural resource management in Malawi and Botswana. *World Development* 34(11): 1942–1957.

Blaut J.M. (1993) *The colonizer's model of the world: geographical diffusionism and Eurocentric history.* New York, USA: The Guilford Press.

Boellstorff T. (2004) Playing back the nation: waria, Indonesian transvestites. *Cultural Anthropology* 19(2): 159–195.

Boellstorff T. (2005) Between religion and desire: being Muslim and gay in Indonesia. *American Anthropologist* 107(4): 575–585.

Bonilla Y. (2020) The coloniality of disaster: race, empire, and the temporal logics of emergency in Puerto Rico, USA. *Political Geography* 78. doi: 10.1016/j. polgeo.2020.102181

Boone J.A. (2015) *The homoerotics of Orientalism.* New York, USA: Colombia University Press.

Borrero J.C. (2005) Field data and satellite imagery of tsunami effects in Banda Aceh. *Science* 308(5728): 1596.

Bourque L.B., Shoaf K.I., Nguyen L.H. (1997) Survey research. *International Journal of Mass Emergencies and Disasters* 15(1): 71–101.

Boyne R. (1990) *Foucault and Derrida: the other side of reason.* London, UK: Routledge.

Bradshaw S. (2013) *Gender, development and disasters.* Cheltenham, UK: Edward Elgar.

Bradshaw S. (2014) Engendering development and disasters. *Disasters* 39(S1): S54–S75.

Buckle P. (2006) Assessing social resilience. In Paton D., Johnston D. (Eds) *Disaster resilience: an integrated approach.* Springfield, USA: Charles C. Thomas Publisher, pp. 88–104.

Burton I., Kates R.W. (1964) The perception of natural hazards in resource management. *Natural Resources Journal* 3(3): 412–441.

Burton I., Kates R.W., White G.F. (1978) *The environment as hazard.* New York, USA: Oxford University Press.

Butler J. (1990) *Gender trouble: feminism and the subversion of identity.* New York, USA: Routledge.

Butler J. (1993) *Bodies that matter: on the discursive limits of 'sex'.* London, UK: Routledge.

Butler J. (1995) Contingent foundations: feminism and the question of "postmodernism". In Benhabib S., Butler J., Cornell D., Fraser N. (Eds) *Feminist contentions: a philosophical exchange.* New York, USA: Routledge, pp. 35–57.

Butler J. (1997) *The psychic life of power: theories in subjection.* Stanford, USA: Stanford University Press.

Butler J. (2004) *Undoing gender.* New York, USA: Routledge.

Butler J. (2016) Rethinking vulnerability and resistance. In Butler J., Gambetti Z., Sabsay L. (Eds) *Vulnerability in resistance.* Durham, USA: Duke University Press, pp. 12–27.

Butler M., Slade G., Dias C.M. (2018) Self-governing prisons: prison gangs in an international perspective. *Trends in Organized Crime.* doi: 10.1007/s12117-018-9338-7

Campbell J.R. (1990) Disasters and development in historical context: tropical cyclone response in the Banks Islands of Northern Vanuatu. *International Journal of Mass Emergencies and Disasters* 8(3): 401–424.

Campbell J.R. (2006) *Traditional disaster reduction in Pacific island communities.* GNS Science Report 2006/038, GNS Science, Lower Hutt, New Zealand.

Canagarajah A.S. (2002) *A geopolitics of academic writing.* Pittsburgh, USA: University of Pittsburgh Press.

Candaliza-Gutierrez F. (2012) Pangkat: inmate gangs at the New Bilibid Prison maximum security compound. *Philippine Sociological Review* 60:193–238.

Canguilhem G. (1966) *Le normal et le pathologique.* Paris, France: Presses Universitaires de France.

Cannon T. (1994) Vulnerability analysis and the explanation of 'natural' disasters. In Varley A. (Ed.) *Disasters, development and environment.* Chichester, UK: J. Wiley & Sons, pp. 13–30.

Cannon T. (2014) The myth of community? In International Federation of Red Cross and Red Crescent Societies (Ed.) *World disaster report 2014: focus on culture and risk.* Geneva, Switzerland: International Federation of Red Cross and Red Crescent Societies, pp. 93–119.

Carabine E. (2015) *Revitalising evidence-based policy for a post-2015 disaster risk reduction framework: lessons from existing international science mechanisms.* London, UK: Overseas Development Institute.

Carbado D.W., Crenshaw K.W., Mays V.M., Tomlinson B. (2013) Intersectionality: mapping the movements of a theory. *Du Bois Review* 10(2): 303–312.

Carlton S., Mills C.E. (2017) The student volunteer army: a 'repeat emergent' emergency response organisation. *Disasters* 41(4): 764–787.

Carrigan A. (2010) Postcolonial disaster, Pacific nuclearization, and disabling environments. *Journal of Literary and Cultural Disability Studies* 4(3): 255–72.

Carrigan A. (2015) Towards a postcolonial disaster studies. In Deloughrey E., Didur J., Carrigan A. (Eds.) *Global ecologies and the environmental humanities.* Abingdon, UK: Routledge, pp. 117–139.

Castel R. (1981) *La gestion des risques.* Paris, France: Editions de Minuit.

Castree N. (2014) *Making sense of nature: representation, politics and democracy.* London, UK: Routledge.

Castree N. (2017) Unfree radicals: geoscientists, the Anthropocene, and left politics. *Antipode* 49(S1): 52–74.

Centre for Research on the Epidemiology of Disasters (2018) *EM-DAT: The International Disaster Database.* Brussels, Belgium: Centre for Research on the Epidemiology of Disasters. http://emdat.be/ (accessed 19 June 2018).

Césaire A. (1950) *Discours sur le colonialisme.* Paris, France: Editions Réclame.

Césaire A. (1956) Culture et colonialisation. *Présence Africaine* 8/10: 190–205.

Chakrabarty D. (1995) Radical histories and question of Enlightenment rationalism: some recent critiques of "Subaltern Studies". *Economic and Political Weekly* 30(14): 751–759.

Chakrabarty D. (2000) *Provincializing Europe: postcolonial thought and historical difference.* Princeton, USA: Princeton University Press.

Chambers E. (1728) *Cyclopædia: or, an universal dictionary of arts and sciences.* Two volumes. London, UK: James and John Knapton et al.

Chambers R. (1974) *Managing rural development: ideas and experience from East Africa.* Uppsala, Sweden: Scandinavian Institute of African Studies.

Chambers R. (1981) Rapid rural appraisal: rationale and repertoire. *Public Administration and Development* 1(2): 95–106.

Chambers R. (1983) *Rural development: putting the last first.* London, UK: Longman.

Chambers R. (1984) Rapid rural appraisal: rationale and repertoire. *Public Administration and Development* 1(2): 65–106.

Chambers R. (1994) Participatory rural appraisal (PRA): analysis of experience. *World Development* 22(9): 1253–1268.

Chambers R. (1995) Paradigm shifts and the practice of participatory research and development. In Nelson N., Wright S. (Eds) *Power and participatory development: theory and practice.* London, UK: Intermediate Technology Publications, pp. 30–42.

Chambers R. (1997) *Whose reality counts? Putting the first last.* London, UK: Intermediate Technology Publications.

Chambers R. (1998) Foreword. In Gujit I., Shah M.K. (Eds) *The myth of community: gender issues in participatory development.* Rugby, UK: Intermediate Technology Publications, pp. xviii–xx.

Chambers R. (2003) Participation and numbers. *PLA Notes* 47: 6–12.

Chambers R. (2007a) *Poverty research: methodologies, mindsets and multidimensionality.* Working Paper No 293. Brighton, UK: Institute of Development Studies.

Chambers R. (2007b) *From PRA to PLA and pluralism: practice and theory.* Working Paper No 286. Brighton, UK: Institute of Development Studies.

Chambers R. (2010) A revolution whose time has come? The win-win of quantitative participatory approaches and methods. *IDS Bulletin* 41(6): 45–55.

Chaplin D., Twigg J., Lovell E. (2019) *Intersectional approaches to vulnerability reduction and resilience-building.* Resilience Intel No. 12. London, UK: Overseas Development Institute.

Chapman D.M. (1994) *Natural hazards.* Melbourne, Australia: Oxford University Press.

Chatterjee P (1983) Peasants, politics and historiography: a response. *Social Scientist* 11(5): 58–65.

Chatterjee P. (2008) Democracy and economic transformation in India. *Economic and Political Weekly* 43(16): 53–62.

Cheek W.W., Chmutina K. (2021) Suppose we had a (resilient) city. *International Journal of Disaster Risk Reduction*: in press.

Chibber V. (2013) *Postcolonial theory and the spectre of capital.* London, UK: Verso.

Chmutina K., Sadler N., von Meding J., Abukhalaf A. (2020) Lost (and found?) in translation: key terminology in disaster studies. *Disaster Prevention and Management* 30(2): 149–162.

Cho S., Crenshaw K.W., McCall L. (2013) Toward a field of intersectionality studies: theory, applications, and praxis. *Signs: Journal of Women in Culture and Society* 38(4): 785–810.

Chomsky N., Foucault M. (2006) *The Chomsky-Foucault debate on human nature.* New York, USA: The New Press.

Christie N. (1993) *Crime control as industry: towards gulags, Western style.* New York, USA: Routledge.

Cianfarani M. (2013) Supporting the resilience of LGBTQI2S people and households in Canadian disaster and emergency management. MA thesis, University of Toronto, Toronto, Canada.

Cicero M.T. (~87 BC, republished 1949) *De inventione. De optimo genere oratorum. Topica.* Cambridge, USA/London, UK: Harvard University Press/William Heinemann Ltd.

Clark N. (2011) *Inhuman nature: sociable life on a dynamic planet.* Los Angeles, USA: Sage.

Cleaver F. (1999) Paradoxes of participation: questioning participatory approaches to development. *Journal of International Development* 11(4): 597–612.

Clemer D. (1940) *The prison community.* New Braunfels, USA: Christopher Publishing House.

Clifford J. (1986a) Introduction: partial truth. In Clifford J., Marcus G.E. (Eds.) *Writing culture: the poetics and politics of ethnography.* Berkeley, USA: University of California Press, pp. 1–26.

Clifford J. (1986b) On ethnographic allegory. In Clifford J., Marcus G.E. (Eds.) *Writing culture: the poetics and politics of ethnography.* Berkeley, USA: University of California Press, pp. 98–121.

Clifford J. (1988) *The predicament of culture: twentieth century ethnography, literature, and art.* Cambridge, USA: Harvard University Press.

Coetzee C., van Niekerk D. (2012) Tracking the evolution of the disaster management cycle: a general system theory approach. *Jàmbá: Journal of Disaster Risk Studies* 4(1): doi: 10.4102/jamba.v4i1.54

Collogan L.K., Tuma F., Dolan-Sewell R., Borja S., Fleischman A.R. (2004) Ethical issues pertaining to research in the aftermath of disaster. *Journal of Traumatic Stress* 17(5): 363–372.

Comité d'Information Sahel (1975) *Qui se nourrit de la famine en Afrique?.* Paris, France: F. Maspero.

Connell R. (2007) *Southern theory: the global dynamics of knowledge in social science.* Crows Nest: Australia: Allen & Unwin.

Constantino R. (1978) *Neocolonial identity and counter-consciousness: essays on cultural decolonization.* White Plains, USA: M.E. Sharpe Inc.

Cooke B., Kothari U. (2001) *Participation: the new tyranny?* London, UK: Zed Books.

Cooke B. (2004) Rules of thumbs for participatory change agents. Hickey S., Mohan G. (Eds) *Participation: from tyranny to transformation.* London, UK: Zed Books, pp. 42–55.

Copans J. (Ed.) (1975) *Sécheresses et famines du Sahel.* Paris, France: F. Maspero.

Cornwall A. (2004) Spaces for transformation? Reflection on issues of power and difference in participation in development. In Hickey S., Mohan G. (Eds) *Participation: from tyranny to transformation.* London, UK: Zed Books, pp. 75–91.

Cornwall A., Brock K. (2005) What do buzzwords do for development policy? A critical look at 'participation', 'empowerment' and 'poverty reduction'. *Third World Quarterly* 26(7): 1043–1060.

Crenshaw K. (1989) Demarginalising the intersection of race and sex: a black feminist critique of antidiscrimination doctrine, feminist theory and antiracist politics. *The University of Chicago Legal Forum* 1989(1): 139–167.

Crenshaw K. (1991) Mapping the margins: intersectionality, identity politics, and violence against women of color. *Stanford Law Review* 43(6): 1241–1299.

Crewe B. (2007) Power, adaptation and resistance in a late-modern men's prison. *British Journal of Criminology* 47(2): 256–275.

Crewe B. (2012) *The prisoner society: power, adaptation and social like in an English prison.* London, UK: Macmillan Publishing.

Crosby A.W. (1997) *The measure of reality: quantification in Western Europe, 1250–1600.* Cambridge, UK: Cambridge University Press.

Crump T. (1990) *The anthropology of numbers.* Cambridge, UK: Cambridge University Press.

Culibar M.J.A.T. (2020) Disaster realities in game world. Master's thesis, The University of Auckland, Auckland, New Zealand.

Cuny F. (1983) *Disasters and development.* Oxford, UK: Oxford University Press.

Cupples J. (2011) Boundary crossings and new striations: when disaster hits a neoliberalising campus. *Transactions of the Institute of British Geographers* 37(3): 337–341.

Cutter S., Burton C.G., Emrich C.T. (2010) Disaster resilience indicators for benchmarking baseline conditions. *Journal of Homeland Security and Emergency Management* 7(1): Art. 51.

D'Ooge, C. (2008) Queer Katrina: gender and sexual orientation matters in the aftermath of the disaster. In Willinger B. (Ed.) *Katrina and the women of New Orleans.* Newcomb College Centre for Research on Women, New Orleans, USA: Tulane University, pp. 22–24.

Dahl R.A. (1957) The concept of power. *Behavioral Science* 2(3): 201–215.

Dalisay S.N.M. (2008) Survival strategies to overcome inaagosto and nordeste in two coastal communities in Batangas and Mindoro, the Philippines. *Disaster Prevention and Management* 17(3): 373–382.

Darke S. (2018) *Conviviality and survival: co-producing Brazilian prison order.* London, UK: Palgrave Macmillan.

Das V. (1989) Subaltern as perspective. In Guha R. (Ed.) *Subaltern Studies VI: Writing on South Asian History and Society.* Delhi, India: Oxford University Press, pp. 310–324.

Davis I. (1978) *Shelter after disaster.* Oxford, UK: Oxford: Polytechnic Press.

Davis I. (1984) Prevention is better than cure. *RRDC Bulletin* October: 3–7.

Davis I., Haghebeart B., Peppiatt D. (2004) *Social vulnerability and capacity analysis.* Discussion paper and workshop report, Geneva, Switzerland: ProVention Consortium.

Dazé A., Ambrose K., Ehrhart C. (2009) *Climate vulnerability and capacity analysis handbook.* Geneva, Switzerland: CARE International.

de Certeau M. (1980) *L'invention du quotidien, 1. Arts de faire.* Paris, France: Flammarion.

de Jaucourt L. (1756) Europe. In Diderot D., d'Alembert J.L.R. (Eds) *Encyclopédie ou dictionnaire raisonné des sciences, des arts et des métiers par une société de gens de lettres.* Paris, France: Briasson / David / Le Breton / Durand, pp. 211–212.

Defoe D. (1704) *The storm: or a collection of the most remarkable casualties and disasters which happen'd in the late dreadful tempest both by sea and land.* London, UK: G. Sawbridge.

Delica Z.G. (1993) Citizenry-based disaster preparedness in the Philippines. *Disasters* 17(3):239–247.

Delica Z.G. (1994) *Vulnerabilities of women and children in disaster situations.* In *Proceedings of the National Conference on Natural Disaster Mitigation,* 19–21 October 1994, Manila. Philippine Institute of Volcanology and Seismology, Quezon City, The Philippines, pp. 258–262.

Delica Z.G. (1999) Community mobilization for early warning. *Philippine Planning Journal* 30(2): 30–40.

Delica-Willison Z., Gaillard J.C. (2012) Community action and disaster. In Wisner B., Gaillard J.C., Kelman I. (Eds.) *Handbook of hazards and disaster risk reduction.* Abingdon, UK: Routledge, pp. 711–722.

Delica-Willison Z., dela Cruz L., Molina F.G.J. (2017) Communities doing disaster risk reduction including climate change adaptation. In Kelman I., Mercer J., Gaillard J.C. (Eds) *The Routledge handbook of disaster risk reduction including climate change adaptation.* Abingdon, UK: Routledge, pp. 340–351.

Delphy C. (1991) Penser le genre: quels problèmes? In Hurtig M.-C., Kail M., Rouch H. (Eds) *Sexe et genre. De la hiérarchie entre les sexes*. Paris, France: Presses du CNRS, pp. 89–101.

Demeritt D. (2001) The construction of global warning and the politics of science. *Annals of the Association of American Geographers* 91(2): 307–337.

Derrida J. (1967a) *De la grammatologie*. Paris, France: Les Editions de Minuit.

Derrida J. (1967b) *L'écriture et la différence*. Paris, France: Editions du Seuil.

Derrida J. (1972) *Marges de la philosophie*. Paris, France: Les Editions de Minuit.

Derrida J. (1981) Les morts de Roland Barthes. *Poétique* 47: 269–292.

Derrida J. (1983) The time of a thesis: punctuations. In Montefiore A. (Ed.) *Philosophy in France today*. Cambridge, UK: Cambridge University Press, pp. 34–50.

Derrida J. (1987) *Psyché: Inventions de l'autre*. Paris, France: Galilée.

Derrida J. (2003) *Psyché: Inventions de l'autre, tome II*. Paris, France: Galilée.

Derrida J. (2004a) Qu'est-ce que la déconstruction? *Commentaire* 108: 1099–1100.

Derrida J. (2004b) Je suis en guerre contre moi-même. *Le Monde* 18 August: https://www.lemonde.fr/archives/article/2004/08/18/jacques-derrida-je-suis-en-guerre-contre-moi-meme_375883_1819218.html

Derrida J., Mortley R. (1991) Jacques Derrida. In Mortley R. (Ed.) *French philosophers in conversation: Levinas, Schneider, Serres, Irigaray, Le Doeuff, Derrida*. London, UK: Routledge, pp. 92–108.

de Sagun J.P. (2021) Mga karanasan ng LGBT San Nicolas sa panahon ng kalamidad. Presentation at the *Online Forum on Climate Change-Related Experiences of Non-Normative Genders and Sexualities and their Responses*. University of the Philippines Resilience Institute, The Philippines, 20 January 2021, Online.

de Sousa Santos B. (Ed.) (2007) *Another knowledge is possible: beyond Northern epistemologies*. London, UK: Verso.

Desai V., Potter R. (2006) *Doing development research*. London, UK: Sage.

Diagne S.B. (2013) On the postcolonial and the universal? *Rue Descartes* 78: 7–18.

Dirlik A. (1997) *The postcolonial aura: Third World criticism in the age of global capitalism*. Boulder, USA: Westview Press.

Dobson N. (1994) From under the mud-pack: women and the Charleville floods. *The Macedon Digest* 9(2): 11–13.

Dominey-Howes D., Gorman-Murray A., McKinnon S. (2014) Queering disasters: on the need to account for LGBTI experiences in natural disaster contexts. *Gender, Place and Culture* 21(7): 905–918.

Donnelly J. (2013) *Universal human rights in theory and practice*. Ithaca, USA: Cornell University Press.

Donner S.D., Webber S. (2014) Obstacles to climate change adaptation decisions: a case study of sea-level rise and coastal protection measures in Kiribati. *Sustainability Science* 9(3): 331–345.

Doron R.O. (2021) Untitled. Presentation at the *Online Forum on Climate Change-Related Experiences of Non-Normative Genders and Sexualities and their Responses*. 20 January 2021, University of the Philippines Resilience Institute, The Philippines, Online.

Drabek T.E. (1986) *Human system responses to disaster: an inventory of sociological findings*. New York, USA: Springer-Verlag.

Dynes R.R. (2000) The dialogue between Voltaire and Rousseau on the Lisbon earthquake: the emergence of a social science view. *International Journal of Mass Emergencies and Disasters* 18(1): 97–115.

Dynes R.R., Haas E., Quarantelli E.L. (1967) Administrative, methodological, and theoretical problems of disaster research. *Indian Sociological Bulletin* 4: 215–227.

Eboussi-Boulaga F. (1977) *La crise du Muntu.* Paris, France: Présence Africaine.

Edwards M., Hulme D. (1995) *Non-governmental organisations – Performance and accountability: beyond the magic bullet.* London, UK: Earthscan.

Effendy B. (2003) *Islam and the state in Indonesia.* Singapore: Institute of Southeast Asian Studies.

Ellwood D.W. (1992) *Rebuilding Europe: Western Europe, American and postwar reconstruction.* London, UK: Routledge.

Enarson E. (1998) Through women's eyes: a gendered research agenda for disaster social science. *Disasters* 22(2): 157–173.

Enarson E., Dhar Chakrabarti P.G. (2009) *Women, gender and disaster: global issues and initiatives.* New Delhi, India: Sage.

Enarson E., Fordham M. (2001) From women's needs to women's rights in disasters. *Environmental Hazards* 3(3): 133–136.

Enarson E., Morrow B. (1998) *The gendered terrain of disaster: through women's eyes.* Westport, USA: Praeger.

Enarson E., Pease B. (2016) *Men, masculinities and disasters.* Abingdon, UK: Routledge.

Enarson E., Phillips B. (2008) Invitation to a new feminist disaster sociology. In Phillips B.D., Morrow B.H. (Eds) *Women and disasters: from theory to practice.* Bloomington, USA: Xlibris, pp. 41–74.

Enriquez V.G. (1992) *From colonial to liberation psychology: the Philippine experience.* Quezon City, The Philippines: University of the Philippines Press.

Escobar A. (1991) Elements of a post-structuralist political ecology. *Futures* 28(4): 325–343.

Escobar A. (1995) *Encountering development: the making and unmaking of the Third World.* Princeton, USA: Princeton University Press.

Fanon F. (1952) *Peau noire, masques blancs.* Paris, France: Editions du Seuil.

Fanon F. (1956) Racisme et culture. *Présence Africaine* 8/10: 122–131.

Fanon F. (1961) *Les damnés de la terre.* Paris, France: F. Maspero.

Felli R., Castree N. (2012) Neoliberalising adaptation to environmental change: foresight or foreclosure? *Environment and Planning A* 44(1): 1–4.

Femia J.V. (1981) *Gramsci's political thought: hegemony, consciousness, and the revolutionary process.* Oxford, UK: Clarendon Press.

Ferguson K.E. (1993) *The man question: visions of subjectivity in feminist theory.* Berkeley, USA: University of California Press.

Few R., Brown K., Tompkins E.L. (2007) Public participation and climate change adaptation: avoiding the illusion of inclusion. *Climate Policy* 7(1): 46–59.

Feyerabend P. (1975) *Against method: outline of an anarchistic theory of knowledge.* London, UK: Verso.

Feyerabend P. (1987) *Farewell to reason.* London, UK: Verso.

Flyvbjerg B. (1998) *Rationality and power: democracy in practice.* Chicago, USA: University of Chicago Press.

Folke C., Carpenter S., Elmqvist T., Gunderson L., Holling C.S., Walker B., Bengtsson J., Berkes F., Colding J., Danell K., Falkenmark M., Gordon L., Kasperson R., Kautsky N., Kinzig A., Levin S., Mäler K.-G., Moberg F., Ohlsson L., Olsson P., Ostrom E., Reid W., Rockström J., Savenije H., Svedin U. (2002) *Resilience and sustainable development: building adaptive capacity in a world of transformations.* Scientific Background Paper on Resilience for the process of The World

Summit on Sustainable Development, Stockholm, Sweden: The Environmental Advisory Council to the Swedish Government.

Fordham M. (1998) Making women visible in disasters: problematising the private domain. *Disasters* 22(2): 126–143.

Fordham M. (2012) Gender, sexuality and disaster. In Wisner B., Gaillard J.C., Kelman I. (Eds) *Handbook of natural hazards and disaster risk reduction*. Abingdon, UK: Routledge, pp. 424–435.

Forthergill A. (1996) The neglect of gender in disaster work: an overview of the literature. *International Journal of Mass Emergencies and Disasters* 14(1): 33–56.

Forthergill A. (1999) Women's roles in a disaster. *Applied Behavioral Science Review* 7(2): 125–143.

Foster H. (1983) *The anti-aesthetic: essays on postmodern culture*. Port Townsend, USA: Bay Press.

Foucault M. (1966) *Les mots et les choses: une archéologie des sciences humaines*. Paris, France: Gallimard.

Foucault M. (1969a) Qu'est-ce qu'un auteur? *Bulletin de la Société Française de Philosophie* 3: 73–104.

Foucault M. (1969b) *L'archéologie du savoir*. Paris, France: Gallimard.

Foucault M. (1971) *L'ordre du discours*. Paris, France: Gallimard.

Foucault M. (1975) *Surveiller et punir: naissance de la prison*. Paris, France: Gallimard.

Foucault M. (1976) *Histoire de la sexualité I: la volonté de savoir*. Paris, France: Gallimard.

Foucault M. (1977a) El poder, una bestia magnifica. *Cuadernos para el Diálogo* 238: 60–63.

Foucault M. (1977b) Les rapports de pouvoir passent à l'intérieur des corps. *La Quinzaine Littéraire* 247: 4–6.

Foucault M. (1978) Introduction. In Canguilhem G. *On the normal and the pathological*. Dordrecht, The Netherlands: D. Reidel Publishing Company, pp. ix–xx.

Foucault M. (1982) The subject and power. *Critical Inquiry* 8(4): 777–795.

Foucault M. (1984a) What is Enlightenment? In Rabinow P. (Ed.) *The Foucault reader*. New York, USA: Pantheon Books, pp. 32–50.

Foucault M. (1984b) An interview: sex, power and the politics of identity. *The Advocate* 400: 26–30, 58.

Foucault M. (1997) *Il faut défendre la société: cours au Collège de France – 1976*. Paris, France: Ecole des Hautes Etudes en Sciences Sociales / Gallimard / Editions du Seuil.

Foucault M. (2004a) *Sécurité, territoire, population: cours au Collège de France – 1977– 1978*. Paris, France: Ecole des Hautes Etudes en Sciences Sociales / Gallimard / Editions du Seuil.

Foucault M. (2004b) *La naissance de la biopolitique: cours au Collège de France – 1978– 1979*. Paris, France: Ecole des Hautes Etudes en Sciences Sociales / Gallimard / Editions du Seuil.

Foucault M., Deleuze G. (1972) Les intellectuels et le pouvoir. *L'Arc* 49: 3–10.

Foucault M., Colas D., Grosrichard A., Le Gaufey G., Livi J., Miller G., Miller J.-A., Millor C., Wajeman G. (1977) Le jeu de Michel Foucault. *Ornicar: Bulletin Périodique du Champ Freudien* 10: 62–93.

Frampton S., McNaught A., Chaffey J., Hardwick J. (2000) *Natural hazards*. 2nd edition. London, UK: Hodder & Stoughton.

Fraser N. (1995) From redistribution to recognition? Dilemmas of justice in a 'post-socialist' age. *New Left Review* 1/212: 68–93.

Fraser N. (2000) Rethinking recognition. *New Left Review* 3: 107–120.

Freire P. (1963) Conscientização e alfabetização uma nova visão do processo. *Revista de Cultura da Universidade do Recife* 4: 5–23.

Freire P. (1968) *Pedagogia do oprimido*. Original manuscript, São Paulo, Brazil: Instituto Paulo Freire.

Freire P. (1970) Cultural action and conscientization. *Harvard Educational Review* 40(3): 452–477.

Fritz C.E. (1961) Disaster. In Merton R.K., Nisbet R.A. (Eds) *Contemporary social problems: an introduction to the sociology of deviant behaviour and social disorganization.* Paris, France, New York, USA: Harcourt, Brace & World, pp. 651–694.

Gaillard J.C. (2015) *People's response to disasters in the Philippines: vulnerability, capacities and resilience.* New York, USA: Palgrave-Macmillan.

Gaillard J.C., Mercer J. (2013) From knowledge to action: bridging gaps in disaster risk reduction. *Progress in Human Geography* 37(1): 93–114.

Gaillard J.C., Navizet F. (2012) Prisons, prisoners and disaster. *International Journal of Disaster Risk Reduction* 1(1): 33–43.

Gaillard J.C., Peek L. (2019) Disaster-zone research needs a code of conduct. *Nature* 575(7783): 440–442.

Gaillard J.C., Cadag J.R.D., Gampell A., Hore K., Le Dé L., McSherry A. (2016a) Participatory numbers for integrating knowledge and actions in development. *Development in Practice* 26(8): 998–1012.

Gaillard J.C., Casing-Baring E.M., Sacayan D., Balay-As M., Santos M. (2016b) *Reducing and managing the risk of disaster in Philippine jails and prisons.* Disaster Prevention and Management Policy Brief Series No 1, Bingley, UK: Emerald.

Gaillard J.C., Gorman-Murray A., Fordham M. (2017a) Sexual and gender minorities in disaster. *Gender, Place and Culture* 24(1): 18–26.

Gaillard J.C., Sanz K., Balgos B.C., Dalisay S.N.M., Gorman-Murray A., Smith F., Toelupe V. (2017b) Beyond men and women: a critical perspective on gender and disaster. *Disasters* 41(3): 429–447.

Gaillard J.C., Cadag J.R.D., Rampengan M.M.F. (2019) People's capacities in facing hazards and disasters: an overview. *Natural Hazards* 95(3): 863–876.

Gampell A.V., Gaillard J.C., Parsons M., Le Dé L. (2020) Exploring the use of the Quake Safe House video game to foster disaster and disaster risk reduction awareness in museum visitors. *International Journal of Disaster Risk Reduction* 49: doi: 10.1016/j.ijdrr.2020.101670

Garcia J.N.C. (2008) *Philippine gay culture: binabae to bakla, silahis to MSM.* Quezon City, The Philippines: University of the Philippines Press.

Garcia-Acosta V. (2009) Presentación. In Ribeiro G.L., Escobar A. (Eds) *Antropologías des mundo: transformaciones disciplinarias dentro de sistemas de poder.* Mexico City, Mexico: Centro de Investigaciones y Estudios Superiores en Antropología Social/Universidad Autónoma Metropolitana/Universidad Iberoamericana/Envión, pp. 17–21.

Gaventa J. (2006) Finding the spaces for change: a power analysis. *IDS Bulletin* 37(6): 23–33.

Gibson T. (2019) *Making aid agencies work: reconnecting INGOs with the people they serve.* Bingley, UK: Emerald Publishing.

Gilbert C. (1995) Studying disaster: a review of the main conceptual tools. *International Journal of Mass Emergencies and Disasters* 13(3): 231–240.

Glissant E. (1981) *Le discours antillais.* Paris, France: Le Seuil.

Global Network of Civil Society Organisations for Disaster Reduction (2013) *Views from the front line 2013.* London, UK: Global Network of Civil Society Organisations for Disaster Reduction.

Gomez C., Hart D.E. (2013) Disaster gold rushes, sophisms and academic neocolonialism: comments on "Earthquake disasters and resilience in the global North". *The Geographical Journal* 179: 272–277.

Goodale M. (2006) Toward a critical anthropology of human rights. *Current Anthropology* 47(3): 485–511.

Gordon J.E. (1978) *Structures.* Harmondsworth, UK: Penguin Books.

Gramsci A. (1930) Alcuni temi della questione meridionale. *Lo Stato Operaio* 4(1): 9–26.

Gramsci A (1929–35) *Quaderni del carcere.* Available from: https://quadernidelcarcere.wordpress.com/

Gramsci A. (1971) *Selections from the prison notebooks.* New York: International Publishers.

Guha R. (1982) On some aspects of the historiography of colonial India. In Guha R. (Ed.) *Subaltern studies I.* New Delhi, India: Oxford University Press, pp. 1–8.

Guha R. (1997) *Dominance without hegemony: history and power in colonial India.* Cambridge, USA: Harvard University Press.

Guha R., Spivak G.C. (1988) *Selected subaltern studies.* New York, USA: Oxford University Press.

Guillermo R. (2009) *Pook at paninindigan: kritika ng pantayong pananaw.* Quezon City, The Philippines: University of the Philippines Press.

Gujit I., Shah M.K. (1998) *The myth of community: gender issues in participatory development.* Rugby, UK: Intermediate Technology Publications.

Gunder M. (2003) Passionate planning for the others' desire: an agonistic response to the dark side of planning. *Progress in Planning* 60(3): 235–319.

Gunewardena N., Schuller M. (2008) *Neoliberal strategies in disaster reconstruction.* Lanham, USA: AltaMira Press.

Haas J.E., Kates R.W., Bowden M.J. (1977) *Reconstruction following disaster.* Cambridge, USA: The MIT Press.

Habermas J. (1985) *Der philosophische diskurs der moderne.* Berlin, Germany: Suhrkamp Verlag.

Hagman G., Beer H., Bendz M., Wijkman A. (1984) *Prevention is better than cure: report on human and environmental disasters in the Third World.* Stockholm, Sweden: Swedish Red Cross.

Hall B.L. (1978) *Creating knowledge: breaking the monopoly – Research methods, participation and development. International Council for Adult Education,* Paris, France: United Nations Educational, Scientific and Cultural Organization.

Hallegatte S., Bangalore M., Jouanjean M.-A. (2016) *Higher losses and slower development in the absence of disaster risk management investments.* Policy Research Working Paper 7632, Washington, USA: The World Bank.

Hardt M., Negri A. (2004) *Multitude: war and democracy in the age of empire.* New York, USA: The Penguin Press.

Hartmann B., Boyce J.K. (1983) *A quiet violence: view from a Bangladesh village.* London, UK: Zed Books.

Harvey D. (1985) *The urbanization of capital.* Oxford, UK: Basil Blackwell.

Harvey D. (1990) *The condition of postmodernity.* Cambridge, UK: Blackwell.

Harvey D. (2005) *A brief history of neoliberalism.* Oxford, UK: Oxford University Press.

Hastrup F. (2011) *Weathering the world: recovery in the wake of the tsunami in a Tamil fishing village.* New York, USA: Berghahn.

Hawley J.C. (2001) *Post-colonial, queer: theoretical intersections.* Albany, USA: SUNY Press.

Hegel G.W.F. (1807) *Phänomenologie des geistes.* Bamberg, Germany: J.A. Goebhardt.

Heijmans A. (2009) The social life of community-based disaster risk management: origins, politics and framing policies. Working paper No. 20, London, UK: Aon Benfield UCL Hazard Research Centre.

Heijmans A. (2012) Risky encounters: institutions and interventions in response to recurrent disasters and conflict. PhD thesis, Wageningen University, Wageningen, The Netherlands.

Heijmans A., Victoria L.P. (2001) *Citizenry-based and development oriented disaster response: experiences and practices in disaster management of the Citizens' Disaster Response Network in the Philippines.* Quezon City, The Philippines: Center for Disaster Preparedness.

Helvarg D. (2010) Interview with a drowning president, Kiribati's Anote Tong. *The Nation* 1 October.

Hermann E., Kempf W. (2017) Climate change and the imagining of migration: emerging discourses on Kiribati's land purchase in Fiji. *The Contemporary Pacific* 2(2): 231–263.

Herdt G. (1994) *Third sex, third gender: beyond sexual dimorphism in culture and history.* Cambridge, USA: Zone Books.

Hewitt K. (1980) The environment as hazard by Ian Burton, Robert W. Kates, and Gilbert F. White. *Annals of the Association of American Geographers* 70(2): 306–311.

Hewitt K. (1983) The idea of calamity in a technocratic age. In Hewitt K. (Ed.) *Interpretations of calamity: from the viewpoint of human ecology.* The Risks and Hazards Series No. 1, Boston, USA: Allen & Unwin Inc, pp. 3–32.

Hewitt K. (1994) Shadow risks and hidden damage: problems in making visible the social space of disasters: from the viewpoint of human ecology. *Proceedings of the Seminario Internacional Sociedad y Prevencion de Desastres*, 23–25 February 1994, Mexico City, Mexico, pp. 1–15.

Hewitt K. (1995a) Sustainable disasters? Perspectives and powers in the discourse of calamity. In Crush J. (Ed.) *Power of development.* London, UK: Routledge, pp. 115–128.

Hewitt K. (1995b) Excluded perspectives in the social construction of disaster. *International Journal of Mass Emergencies and Disasters* 13(3): 317–339.

Hewitt K. (2007) Preventable disasters: addressing social vulnerability, institutional risk, and civil ethics. *Geographische Rundschau International Edition* 3(1): 43–52.

Hilhorst D. (2004) Complexity and diversity: unlocking social domains of disaster response. In Bankoff G., Frerks G., Hilhorst D. (Eds) *Mapping vulnerability: disasters, development and people.* London, UK: Earthscan, pp. 52–66.

Holland J. (2013) *Who counts? The power of participatory statistics.* Rugby, UK: Practical Action Publishing.

Holling C.S. (1973) Resilience and stability of ecological systems. *Annual Review of Ecology and Systematics* 4: 1–23.

Hollnsteiner M.R. (1963) *The dynamics of power in a Philippine municipality.* Quezon City, The Philippines: Institute of Philippine Culture.

Hoquet T. (2016) *Des sexes innombrables: le genre à l'épreuve de la biologie.* Paris, France: Editions du Seuil.

Hore K., Gaillard J.C., Davies T., Kearns R. (2020) People's participation in disaster risk reduction: recentering power. *Natural Hazards Review* 21(2): 10.1061/(ASCE)NH.1527-6996.0000353.

Horkheimer M. (1947) *Eclipse of reason.* Oxford, UK: Oxford University Press.

Horkheimer M., Adorno T. (1947) *Dialektik der Aufklärung.* Amsterdam, The Netherlands: Querido Verlag.

Hoskins T.K., Jones A. (2017) *Critical conversations in kaupapa Māori.* Wellington, New Zealand: Hui Press.

Hountondji P.J. (1976) *Sur la "philosophie africaine".* Paris, France: Maspero.

Hountondji P.J. (2009) Knowledge of Africa, knowledge by Africans: two perspectives on African studies. *RCCS Annual Review* 1: 121–131.

Human Rights Watch (2005) *New Orleans: prisoners abandoned to floodwaters.* New York, USA: Human Rights Watch. Available from: http://www.hrw.org/news/2005/09/21/new-orleans-prisoners-abandoned-floodwaters (accessed 8 January 2021).

Hurtig M.-C., Pichevin, M.-F. (1985) La variable sexe en psychologie: donné ou construct? *Cahiers de Psychologie Cognitive* 5(2): 187–228.

Husserl E. (1954) *Die Krisis der europäischen Wissenschaften und die transzendentale Phänomenologie: eine Einleitung in die phänomenologische Philosophie.* The Hague, The Netherlands: Martinus Nijhoff.

Iemura H., Takahashi Y., Pradono M.H., Sukamdo P., Kurniawan R. (2006) Earthquake and tsunami questionnaires in Banda Aceh and surrounding areas. *Disaster Prevention and Management* 15(1): 21–30.

Ignatieff M. (2003) *Human rights as politics and idolatry.* Princeton, USA: Princeton University Press.

Ileto R.C. (1979) *Pasyon and revolution: popular movements in the Philippines, 1840–1910.* Quezon City, The Philippines: Ateneo de Manila University Press.

Intergovernmental Oceanographic Commission (2020) *Preparing for community tsunami evacuations: from inundation to evacuation maps, response plans, and exercises.* Manual and Guides 82, Paris, France: Intergovernmental Oceanographic Commission.

International Federation of Red Cross and Red Crescent Societies (2006) *World disaster report 2006: focus on neglected crises.* Geneva, Switzerland: International Federation of Red Cross and Red Crescent Societies.

International Federation of Red Cross and Red Crescent Societies (2007) *VCA toolbox with reference sheets.* Geneva, Switzerland: International Federation of Red Cross and Red Crescent Societies.

International Federation of Red Cross and Red Crescent Societies (2012) *Key determinants of a successful CBDRR programme.* Geneva, Switzerland: International Federation of Red Cross and Red Crescent Societies.

International Gay and Lesbian Human Rights Commission (The) (2011) *The impact of the earthquake, and relief and recovery programs on Haitian LGBT people.* New York, USA: The International Gay and Lesbian Human Rights Commission.

Intergovernmental Panel on Climate Change (2014a) *Climate change 2014: impacts, adaptation, and vulnerability. Part A: global and sectoral aspects. Contribution of Working Group II to the Fifth Assessment Report of the Intergovernmental Panel on Climate Change.* Cambridge, UK: Cambridge University Press.

Intergovernmental Panel on Climate Change (2014b) *Climate change 2014: synthesis report summary for policymakers.* Cambridge, UK: Cambridge University Press.

Irwin J., Cressey D.R. (1962) Thieves, convicts and the inmate culture. *Social Problems* 10(2): 142–155.

JanMohamed A.R. (1985) The economy of Manichean allegory: the function of racial difference in colonialist literature. *Critical Inquiry* 12(1): 59–87.

Jigyasu R. (2005) Disaster: a "reality" or "construct"? Perspective from the "East". In Perry R.W., Quarantelli E.L. (Eds) *What is a disaster? New answers to old questions.* Bloomington, USA: Xlibris, pp. 49–59.

Jocano F.L. (2008) *Filipino value system: a cultural definition.* Quezon City, The Philippines: Punlad Research House.

Jones E. (1981) *The European miracle: environments, economies and geopolitics in the history of Europe and Asia.* Cambridge, UK: Cambridge University Press.

Kafi S. (1992) *Disasters and destitute women: twelve case studies.* Dhaka, Bangladesh: Bangladesh Development Partnership Centre.

Kant I. (1755) *Allgemeine Naturgeschichte und Theorie des Himmels.* Königsberg, Germany: Johann Friederich Petersen.

Kant I. (1784) Was ist Aufklarung? *Berlinischen Monatsschrift* December: 481–94.

Kant I. (1786) Mutmaßlicher Anfang der Menschengeschichte. *Berlinischen Monatsschrift* January: 1–27.

Kates R.W. (1971) Natural hazard in human ecological perspective: hypotheses and models. *Economic Geography* 47(3): 438–451.

Kawata Y., Tsuji Y., Sugimoto Y., Hayashi H., Matsutomi H., Okamura Y., Hayashi I., Kayane H., Tanioka Y., Fujima K., Imamura F., Matsuyama M., Takahashi T., Maki N., Koshimura S., Yasuda T., Shigihara Y., Nishimura Y., Horie K., Nishi Y., Yamamoto H., Fukao J., Seiko S., Takakuwa F. (2005) *Comprehensive analysis of the damage and its impact on coastal zones by the 2004 Indian Ocean tsunami disaster.* Kyoto, Japan: Research Group on the December 26, 2004 Earthquake Tsunami Disaster of Indian Ocean.

Kelly C. (1998–99) Simplifying disasters: developing a model for complex non-linear events. *Australian Journal of Emergency Management* 14(1): 25–27.

Kelman I. (2005) Operational ethics for disaster research. *International Journal of Mass Emergencies and Disasters* 23(3): 141–158.

Kelman I., Gaillard J.C. (2010) Embedding climate change adaptation within disaster risk reduction. In Shaw R., Pulhin J., Pereira J. (Eds) *Climate change adaptation and disaster risk reduction: issues and challenges.* Bradford, UK: Emerald, pp. 23–46.

Kelman I., Karnes E. (2007) Relocalising disaster risk reduction in Boulder. Colorado. *Australian Journal of Emergency Management* 22(1): 18–25.

Kemp R.L. (2007) Assessing the vulnerability of buildings. *Disaster Prevention and Management* 16(4): 611–618.

Kendra J., Wachtendorf T. (2020) Disaster-zone research: no need for a customized code of conduct. *Nature* 578(7795): 363.

Kenney C., Phibbs S. (2015) A Māori love story: community-led disaster management in response to the Ōtautahi (Christchurch) earthquakes as a framework for action. *International Journal of Disaster Risk Reduction* 14(1): 46–55.

Khondker H.H. (1996) Women and floods in Bangladesh. *International Journal of Mass Emergencies and Disasters* 14(3): 281–292.

Killian L.M. (1956) *An introduction to methodological problems of field studies in disasters.* Disaster Study Number 8, Committee on Disaster Studies, Washington, DC, USA: National Academy of Science – National Research Council.

King D.N., Goff J.R. (2010) Benefitting from differences in knowledge, practice and belief: Māori oral traditions and natural hazards science. *Natural Hazards and Earth System Sciences* 10: 1927–1940.

Kiribati Adaptation Project (2011) *Kiribati Adaptation Programme: to reduce Kiribati's vulnerability to climate change, climate variability and sea level rise through adaptation.* Tarawa, Kiribati: Kiribati Adaptation Project.

Knight K., Sollom R. (2012) Making disaster risk reduction and relief programmes LGBTI inclusive: examples from Nepal. *Humanitarian Exchange* 55: 36–39.

Knudson K.E. (1981) Adaptational persistence in the Gilbert Islands. In Force R.W., Bishop B. (Eds) *Persistence and exchange.* Honolulu, USA: Pacific Science Association, pp. 91–99.

Kortschak I. (2010) *Invisible people: poverty and empowerment in Indonesia.* Jakarta, Indonesia: The Lontar Foundation.

Kothari U. (2001) Power, knowledge and social control in participatory development. In Cooke B., Kothari U. (Eds) *Participation: the new tyranny?* London, UK: Zed Books, pp. 139–153.

Kreager D.A., Kruttschnitt C. (2018) Inmate society in the era of mass incarceration. *Annual Review of Criminology* 1: 261–283.

Kuruppu N. (2009) Adapting water resources to climate change in Kiribati: the importance of cultural values and meanings. *Environmental Science and Policy* 12(7): 799–808.

Kuruppu N., Liverman D. (2011) Mental preparation for climate adaptation: the role of cognition and culture in enhancing adaptive capacity of water management in Kiribati. *Global Environmental Change* 21(2): 657–669.

Laclau E., Mouffe C. (2001) *Hegemony and socialist strategy: towards a radical democratic politics.* 2nd edition, London, UK: Verso.

Lakhina S.J. (2020) Co-learning disaster resilience: a person-centred approach to understanding experiences of refuge and practices of safety. PhD thesis, University of Wollongong, Wollongong, Australia.

Latour B. (1987) *Science in action: how to follow scientists and engineers through society.* Cambridge, USA: Harvard University Press.

Latour B., Woolgar S. (1979) *Laboratory life: the construction of scientific facts.* Princeton, USA: Princeton University Press.

Lavell A. (2000) Desastres durante una década: lecciones y avances conceptuales y prácticos en América Latina (1990–1999). *Anuario Política y Social de América Latina* 3: 1–34.

Lavell A., Maskrey A. (2014) The future of disaster risk management. *Environmental Hazards* 13(4): 267–280.

Lavell A., Wisner B., Gaillard J.C., Saunders W. (2012) National planning and disaster. In Wisner B., Gaillard J.C., Kelman I. (Eds) *Handbook of hazards and disaster risk reduction.* Abingdon, UK: Routledge, pp. 617–628.

Lavigne F., Paris R. (2011) *Tsunarisque: le tsunami du 26 décembre 2004 à Aceh, Indonésie.* Paris, France: Publications de la Sorbonne.

Lazreg M. (1994) *The Eloquence of silence: Algerian women in question.* London, UK: Routledge.

Lazreg M. (2017) *Foucault's Orient.* New York, USA: Berghahn Books.

Leal P.A. (2007) Participation: the ascendancy of a buzzword in the neo-liberal era. *Development in Practice* 17(4–5): 539–548.

Le Dé L., Gaillard J.C. (2017) Disaster risk reduction and emergency management in prison: a scoping study from New Zealand. *Journal of Contingencies and Crisis Management* 25(4): 376–381.

Leoni B. (2012) *Migration not a matter of choice but survival, says Kiribati President.* Geneva, Switzerland: United Nations International Strategy for Disaster Reduction. Available from: http://www.unisdr.org/archive/25649.

Lévi-Strauss C. (1954) Les mathématiques de l'homme. *Bulletin International des Sciences Sociales* 6(4): 643–653.

Lévi-Strauss C. (1962) *La pensée sauvage.* Paris, France: Plon.

Levinas E. (1972) *Humanisme de l'autre homme.* Montpellier, France: Fata Morgana.

Levinas E. (1987) *Hors sujet.* Montpellier, France: Fata Morgana.

Levine S. (2014) *Assessing resilience: why quantification misses the point.* Working Paper, London, UK: Humanitarian Policy Group, Overseas Development Institute.

Lewis J. (1976a) The precautionary planning for natural disaster. *Foresight* 2(2): 7–10.

Lewis J. (1976b) New directions: are buildings the answer? *The Architect* 9: 48–50.

Lewis J. (1979) The vulnerable state: an alternative view. In Stephens L.H., Green S.J. (Eds) *Disaster assistance: appraisal, reform and new approaches.* London, UK: The Macmillan Press, pp. 104–129.

Lewis J. (1999) *Development in disaster-prone places: studies in vulnerability.* London, UK: IT Publications.

Lewis J. (2003) Housing construction in earthquake-prone places: perspectives, priorities and projections for development. *Australian Journal of Emergency Management* 18(2): 35–44.

Lewis J. (2008) The worm in the bud: corruption, construction, and catastrophe. In Bosher L. (Ed.) *Hazards and the built environment: attaining built-in resilience.* London, UK: Routledge, pp. 238–263.

Lorde A. (1984) *Sister outsider.* Berkeley, USA: Ten Speed Press.

Lugones M. (2007) Heterosexualism and the colonial/modern gender system. *Hypatia* 22(1): 186–209.

Lugones M. (2010) Toward a decolonial feminism. *Hypatia* 25(4): 742–759.

Lukes S. (2005) *Power: a radical view.* 2nd edition, London, UK: Red Globe Press,

Luna E.M. (2001) Disaster mitigation and preparedness: the case of NGOs in the Philippines. *Disasters* 25(3): 216–226.

Luna E.M. (2003) Endogenous system of responses to river flooding as a disaster subculture: a case study of Bula, Camarines Sur. *Philippine Sociological Review* 51: 135–153.

Lundsgaarde H.P. (1966) *Cultural adaptations in the Gilbert Islands.* Department of Anthropology, Eugene, USA: University of Oregon.

Lundsgaarde H.P. (1967) *Social changes in the southern Gilbert Islands, 1938–1964.* Department of Anthropology, Eugene, USA: University of Oregon.

Lyotard J.F. (1979) *La condition postmoderne: rapport sur le savoir.* Paris, France: Les Editions de Minuit.

Lyotard J.F. (1993) The other's rights. In Shute S., Hurley S. (Eds) *On human rights.* New York, USA: Basic Books, pp. 135–147.

MacKenzie U.N. (2004) *'The sun has come down closer to my island': people's perceptions of climate change.* Tarawa, Kiribati: Kiribati Adaptation Project.

Mahdavifar M.R., Izadkhah Y.O., Heshmati V. (2009) Appropriate and correct reactions during earthquakes: 'drop, cover and hold on' or 'triangle of life'. *Journal of Seismology and Earthquake Engineering* 11(1): 42–48.

Mallin M.-A.F. (2018) From see-level rise to seabed grabbing: the political economy of climate change in Kiribati. *Marine Policy* 97: 244–252.

Malm A. (2018) *The progress of this storm: nature and society in a warming world.* London, UK: Verso.

Manalansan M.F. (2006) *Global divas: Filipino gay men in the diaspora.* Quezon City, The Philippines: Ateneo de Manila University Press.

Manyena S.B. (2006) The concept of resilience revisited. *Disasters* 30(4): 433–450.

Manyunga J.S. (2009) Measuring the measure: a multi-dimensional scale model to measure community disaster resilience in the U.S. Gulf Coast region. PhD dissertation, Texas A&M University, College Station, USA.

Mansuri G., Rao V. (2013) *Localizing development: does participation work?* Washington, DC, USA: The World Bank.

Marchezini V. (2015) The biopolitics of disaster: power, discourses, and practices. *Human Organization* 74(4): 363–371.

Marcus G.E., Fischer M.M.J. (1986) *Anthropology as cultural critique: an experimental moment in the human sciences.* Chicago: USA: The University of Chicago Press.

Martin-Baró I. (1994) *Writings for a liberation psychology*. Cambridge, USA: Harvard University Press.

Maskrey A. (1984) Community based hazard mitigation. In *Proceedings of the International Conference on Disaster Mitigation Program Implementation*, 12–16 November 1984, Ocho Rios, Jamaica, pp. 25–39.

Maskrey A. (1989) *Disaster mitigation: a community based approach*. Development Guidelines No 3. Oxford, UK: OXFAM.

Maskrey A. (2011) Revisiting community-based disaster risk management. *Environmental Hazards* 10(1): 42–52.

Maude H.E. (1980) *The Gilbertese maneaba*. Suva, Fiji: University of the South Pacific.

Mayo M. (1975) Community development: a radical alternative? In Bailey R., Brake M. (Eds) *Radical social work*. London, UK: E. Arnold, pp. 129–157.

Mbembe A. (1990) Pouvoir, violence et accumulation. *Politique Africaine* 39: 7–24.

Mbembe A. (1992) Provisional notes on the postcolony. *Africa* 62(1): 3–37.

Mbembe A. (2010) Faut-il provincialiser la France? *Politique Africaine* 119: 159–188.

Mbembe A. (2013) *Critique de la raison nègre*. Paris, France: La Découverte.

Mbembe A. (2016) Decolonizing the university: new directions. *Arts & Humanities in Higher Education* 15(1): 29–45.

McAdoo B.G., Dengler L., Prasetya G., Titov V. (2006) *Smong*: how an oral history saved thousands on Indonesia's Simeulue Island during the December 2004 and March 2005 tsunamis. *Earthquake Spectra* 22(S3): S661–S669.

McEntire D. (2001) Triggering agents, vulnerabilities and disaster reduction: towards a holistic paradigm. *Disaster Prevention and Management* 10(3): 189–196.

McKee N. (1994) *Visualisation in participatory programmes: a manual for facilitators and trainers involved in participatory group events*. Dhaka, Bangladesh: UNICEF Bangladesh.

Mead M. (1949) *Male and female: a study of the sexes in a changing world*. New York, USA: W. Morrow.

Mead M. (1961) Cultural determinants of sexual behavior. In W.C. Young (Ed.) *Sex and internal secretions*. 3rd edition, Baltimore, USA: The Williams & Wilkins Co., pp. 1433–1479.

Memmi A. (1957) *Portrait du colonisé, portrait du colonisateur*. Paris, France: Gallimard.

Menzies S. (2009) *Proposed behaviour change campaign in support of KAP-II – Inception report*. Tarawa, Kiribati: Kiribati Adaptation Project.

Mercer J., Kelman I., Suchet-Pearson S., Lloyd K. (2009) Integrating indigenous and scientific knowledge bases for disaster risk reduction in Papua New Guinea. *Geografiska Annaler B* 91: 157–183.

Mercer J., Kelman I., Taranis L., Suchet-Pearson S. (2010) Framework for integrating indigenous and scientific knowledge for disaster risk reduction. *Disasters* 34: 214–239.

Miyazawa K. (2018) Becoming an insider and an outsider in post-disaster Fukushima. *Harvard Educational Review* 88(3): 334–354.

Mignolo W.D. (2011) *The darker side of Western modernity: global futures, decolonial options*. Durham, USA: Duke University Press.

Mileti D.S. (1987) Sociological methods and disaster research. In Dynes R.R., de Marchi B., Pelanda C. (Eds) *Sociology of disasters: contribution of sociology to disaster research*. Milan, Italy: Franco Angeli, pp. 57–69.

Mishra P. (2009) Let's share the stage: involving men in gender equality and disaster risk reduction. In Enarson E., Dhar Chakrabarti P.G. (Eds) *Women, gender and disaster: global issues and initiatives*. New Delhi, India: Sage, pp. 29–9.

Missbach A. (2011) Ransacking the field? Collaboration and competition between local and foreign researchers in Aceh. *Critical Asian Studies* 43(3): 373–398.

Mohanty C.T. (1984) Under Western eyes: feminist scholarship and colonial discourses. *Boundary* 2 12(3)–13(1): 333–358.

Mohanty C.T. (2003) *Feminism without borders: decolonizing theory, practicing solidarity.* Durham, USA: Duke University Press.

Moncrieffe J., Eyben R. (2007) *The power of labelling: how people are categorized and why it matters.* London, UK: Earthscan.

Moore A., Nishimura Y., Gelfenbaum G., Kamataki T., Triyono R. (2006) Sedimentary deposits of the 26 December 2004 tsunami on the northwest coast of Aceh, Indonesia. *Earth Planets Space* 58: 253–258

Morriss P. (2002) *Power: a philosophical analysis.* Manchester, UK: Manchester University Press.

Moser C.O. (1989) Community participation in urban projects in the Third World. *Progress in Planning* 32(2): 71–133.

Mrazek R. (2002) *Engineers of happy land: technology and nationalism in a colony.* Princeton, USA: Princeton University Press.

Mudimbe V.Y. (1988) *The invention of Africa: gnosis, philosophy, and the order of knowledge.* Bloomington, USA: Indiana University Press.

Narag R.E. (2005) *Freedom and death inside the jail: a look into the condition of the Quezon City Jail.* Manila/Makati, The Philippines: Supreme Court of the Philippines/ United Nations Development Programme.

Narag R.E., Jones C.R. (2017) Understanding prison management in the Philippines: a case for shared governance. *The Prison Journal* 97(1):3–26.

Narag R.E., Jones C.R. (2020) The *kubol* effect: shared governance and cell dynamics in an overcrowded prison system in the Philippines. In Turner J., Knight V. (Eds) *The prison cell.* New York, USA: Palgrave Macmillan, pp. 71–94.

Narag R.E., Lee S. (2018) Putting out fires: understanding the developmental nature and roles of inmate gangs in the Philippine overcrowded jails. *International Journal of Offender Therapy and Comparative Criminology* 62(11): 3509–3535.

National Security, Military, and Intelligence Panel on Climate Change (The) (2020) *A security threat assessment of global climate change: how likely warming scenarios indicate a catastrophic security future.* Washington, DC, USA: The Center for Climate and Security.

Neal D.M. (1997) Reconsidering the phases of disaster. *International Journal of Mass Emergencies and Disasters* 15(2): 239–264.

Nielsen J.M. (1984) *Sex and gender in disaster research.* Department of Sociology, Boulder, USA: University of Colorado Bouder.

Norris F.H., Stevens S.P., Pfefferbaum B., Wyche K.F., Pfefferbaum, R.L. (2008) Community resilience as a metaphor, theory, set of capacities, and strategy for disaster readiness. *American Journal of Community Psychology* 41(1): 127–150.

Norton J., Gibson T.D. (2019) Introduction to disaster prevention: doing it differently by rethinking the nature of knowledge and learning. *Disaster Prevention and Management* 28(1): 2–5.

Norwegian Nobel Committee (2007) *The Nobel peace prize for 2007.* Stockholm, Sweden: Norwegian Nobel Committee. Available from: https://www.nobelprize. org/prizes/peace/2007/press-release/ (accessed 22 February 2021).

Nunn P. (2007) *Climate, environment and society in the Pacific during the last millennium.* Amsterdam, The Netherlands: Elsevier.

O'Brien G., O'Keefe P., Rose J., Wisner B. (2006) Climate change and disaster management. *Disasters* 30(1): 64–80.

O'Hanlon R., Washbrook D. (1992) After Orientalism: culture, criticism, and politics in the Third World. *Comparative Studies in Society and History* 34(1): 141–167.

O'Keefe P., Westgate K., Wisner B. (1976) Taking the naturalness out of natural disasters. *Nature* 260(5552): 566–567.

O'Mathúna D.P. (2010) Conducting research in the aftermath of disasters: ethical considerations. *Journal of Evidence-Based Medicine* 3: 65–75.

Oliver-Smith A. (1996) Anthropological research on hazards and disasters. *Annual Review of Anthropology* 25: 303–328.

Ong J.C. (2017) Queer cosmopolitanism in the disaster zone: 'my Grindr became the United Nations'. *The International Communication Gazette* 79(6–7): 656–673.

Orejas E. (2003) Prevailing over disasters through people's organized action: a continuing engagement in community-based disaster management in Central Luzon. *Philippine Sociological Review* 51: 81–94.

Ortner S.B. (1972) Is female to male as nature is to culture? *Feminist Studies* 1(2): 5–31.

Ostadtaghizadeh A., Ardalan A., Paton D., Jabbari H., Khankeh H.R. (2015) Community disaster resilience: a systematic review on assessment models and tools. *PLOS Currents Disasters* 1. doi: 10.1371/currents.dis.f224ef8efbdfcf1d508dd0de4d8210ed

Oulahen G., Vogel B., Gouett-Hanna C. (2020) Quick response disaster research: opportunities and challenges for a new funding program. *International Journal of Disaster Risk Science* 11(5): 568–577.

OXFAM GB (2010) *Gender, disaster risk reduction, and climate change adaptation: a learning companion.* Oxford, UK: OXFAM GB.

Oyěwùmí O. (1997) *The invention of women: making an African sense of Western gender discourses.* Minneapolis, USA: University of Minnesota Press.

Oyěwùmí O. (2002) Conceptualising gender: the Eurocentric foundations of feminist concepts and the challenge of African epistemologies. *Jenda: A Journal of Culture and African Women Studies* 2(1): 1–7.

Ozawa K. (2012) Relief activities for LGBTI people in the affected areas. *Voices from Japan* 26: 21–22.

Pacific Islands Applied Geoscience Commission (2009) *Guide to developing national action plans: a tool for mainstreaming disaster risk management based on experiences from selected Pacific Island Countries.* Suva, Fiji: Pacific Islands Applied Geoscience Commission.

Padilla-Goodman A. (2010) When "you" become one of "them": understanding the researcher's identity dialectically. *International Review of Qualitative Research* 3(3): 315–330.

Parry B. (1987) Problems in current theories of colonial discourse. *Oxford Literary Review* 9(1/2): 27–58.

Pe-Pua R. (1982) *Sikolohiyang Pilipino: teorya, metodo, at gamit.* Quezon City, The Philippines: University of the Philippines Press.

Pelling M. (2003) *The vulnerabilities of cities: natural disasters and social resilience.* London, UK: Earthscan.

Perlman J.E. (1976) *The myth of marginality: urban poverty and politics in Rio de Janeiro.* Berkeley, USA: University of California Press.

Perry R.W. (1994) A model of evacuation compliance behavior. In Dynes R.R. (Ed.) *Disasters, collective behavior, and social organization.* Newark, USA: University of Delaware Press, pp. 85–98.

Perry R.W., Quarantelli E.L. (2005) *What is a disaster? New answers to old questions.* Bloomington, USA: XLibris.

Petal M., Ronan K., Ovington G., Tofa M. (2020) Child-centred risk reduction and school safety: an evidence-based practice framework and roadmap. *International Journal of Disaster Risk Reduction* 49: 101633.

Philippine Health Research Ethics Board (2011) *National ethical guidelines for health research.* Taguig, The Philippines: Philippine National Health Research System.

Phillips B., Morrow B.H. (2008) *Women and disasters: from theory to practice.* Bloomington, USA: Xlibris Publications.

Pincha C., Krishna H. (2008) Aravanis: voiceless victims of the tsunami. *Humanitarian Exchange* 41: 41–43.

Postema M., Cavallaro J., Nagra R. (2017) Advancing security and human rights by the controlled organization of inmates. *Prison Service Journal* 229: 57–62.

Prakash G. (1994) Subaltern studies as postcolonial criticism. *The American Historical Review* 99(5): 1475–90.

Prashad V. (1993) *The poorer nations: a possible history of the global South.* London, UK: Verso.

Prince S.H. (1920) Catastrophe and social change based upon a sociological study of the Halifax disaster. PhD thesis, Columbia University, New York, USA.

Protevi J. (2006) Katrina. *Symposium* 10(1): 363–381

Protevi J. (2009) *The political affect.* Minneapolis, USA: University of Minnesota Press.

Purdum J.C., Meyer M.A. (2020) Prisoner labor throughout the life cycle of disasters. *Risks, Hazards and Crisis in Public Policy* 11(3): 296–319.

Quarantelli E.L. (1980a) *Sociology and social psychology of disasters: implications for third world and developing countries.* Preliminary Paper No. 66, Disaster Research Center, Newark, USA: University of Delaware.

Quarantelli E.L. (1980b) *The study of disaster movies: research problems, findings and implications.* Preliminary Paper No. 64, Disaster Research Center, Newark, USA: University of Delaware.

Quarantelli E.L. (1981) Disaster planning: small and large – Past, present, and future. In *Proceedings of the American Red Cross EFO Division Disaster Conference,* 19–22 February 1981, Alexandria. Alexandria, USA: American Red Cross Eastern Field Office, pp. 1–26.

Quarantelli E.L. (1990) *Some aspects of disaster planning in developing countries.* Preliminary Paper No. 144, Disaster Research Center, Newark, USA: University of Delaware.

Quarantelli E.L. (1997) The Disaster Center field studies of organized behavior in the crisis time period of disasters. *International Journal of Mass Emergencies and Disasters* 15(1): 47–69.

Quarantelli E.L. (1998) *What is a disaster? A dozen perspectives on the question.* London: Routledge.

Quarantelli E.L., Dynes R.R. (1972) When disaster strikes: it isn't much like what you've heard & read about. *Psychology Today* 5(9): 66–70.

Quarantelli E.L., Davis I. (2011) *An exploratory research agenda for studying the popular culture of disasters (PCD): its characteristics, conditions, and consequences.* Disaster Research Center, Newark, USA: University of Delaware.

Quijano A. (1992) Colonialidad y modernidad/racionalidad. *Perú Indígena* 13(29), 11–20.

Quijano A. (2000a) Coloniality of power and Eurocentrism in Latin America. *International Sociology* 15(2): 215–232.

Quijano A. (2000b) Colonialidad del poder y clasificacion social. *Journal of World-Systems Research* 1(2): 342–386.

Rabinow P. (1977) *Reflections on fieldwork in Morocco.* Berkeley, USA: University of California Press.

Rabinow P. (1996) *Essays on the anthropology of reason.* Princeton, USA: Princeton University Press.

Radcliffe S. (1994) (Representing) Post-colonial women: authority, difference and feminisms. *Area* 26(1): 25–32

Rafael V.L. (1988) *Contracting colonialism: translation and Christian conversion in Tagalog society under early Spanish rule.* Quezon City, The Philippines: Ateneo de Manila University Press.

Rainbird P. (2004) *Archaeology of Micronesia.* Cambridge, UK: Cambridge University Press.

Rancière J. (1987) *Le maître ignorant: cinq leçons sur l'émancipation intellectuelle.* Paris, France: Fayard.

Rappaport J. (1984) Studies in empowerment: introduction to the issue. *Prevention & Intervention in the Community* 3(2–3): 1–7.

Rastogi P. (2020) *Postcolonial disaster: narrating catastrophe in the twenty-first century.* Evanston, USA: Northwestern University Press.

Republic of Kiribati (2006) *Local government act.* Tarawa, USA: Republic of Kiribati.

Republic of Kiribati (2010) *National disaster risk management plan.* Tarawa, USA: Republic of Kiribati.

Resilience Development Initiative (2021) *Komunitas transpuan dan adaptasi perubahan iklim.* Three documentary film series, Bandung, Indonesia: Resilience Development Initiative.

Revet S. (2007) *Anthropologie d'une catastrophe: les coulées de boue de 1999 au Venezuela.* Paris, France: Presse Sorbonne Nouvelle.

Revet S. (2018) *Disasterland: an ethnography of the international disaster community.* New York, USA: Palgrave Macmillan.

Richards P. (Ed.) (1975) *African environment: problems and perspectives.* London, UK: International African Institute.

Ride A., Bretherton D. (2011) *Community resilience in natural disasters.* New York, USA: Palgrave Macmillan.

Rigg J., Oven K. (2015) Building liberal resilience? A critical review from developing rural Asia. *Global Environmental Change* 32: 175–186.

Rivers J.P.W. (1982) Women and children last: an essay on sex discrimination in disasters. *Disasters* 6(4): 256–267.

Robinson-Pant A. (1996) PRA: a new literacy? *Journal of International Development* 8(4): 531–551.

Rousseau J.J. (1967) *Correspondance complète de Jean-Jacques Rousseau. Tome IV – 1756–1757.* Geneva, Switzerland: Les Délices.

Rousseau J.J. (1781, republished 1969) *Essai sur l'origine des langues où il est parlé de la mélodie et de l'imitation musicale.* Paris, France: A.G. Nizet.

Rowlands J. (1995) Empowerment examined. *Development in Practice* 5(2): 101–107.

Rueda-Acosta P.V. (2015) Examining deaths behind bars: towards penal system policy reforms in the context of human rights. PhD thesis, University of the Philippines Diliman, Quezon City, The Philippines.

Safaya R. (1976) *Indian psychology: a critical and historical analysis of the psychological speculations in Indian philosophical literature.* New Delhi, India: Munshiram Manoharlal.

Said E. (1978) *Orientalism.* London, UK: Routledge & Kegan Paul.

Said E. (1985) Orientalism reconsidered. *Cultural Critique* 1: 89–107.

Said E. (1994) *Culture and imperialism*. London, UK: Vintage.

Salazar Z.A. (1991) Ang pantayong pananaw bilang diskursong pangkabihasnan. In Bautista V.V., Pe-Pua R. (Eds) *Pilipinolohiya: kasaysayan, pilosopiya at pananaliksik*. Manila, The Philippines: Kalikasan Press, pp. 46–72.

San Juan E. (1998) *Beyond postcolonial theory*. New York, USA: Palgrave Macmillan.

Scanlon J. (2007) Unwelcome irritant or useful ally? The mass media in emergencies. In Rodríguez H., Quarantelli E.L., Dynes R.R. (Eds) *Handbook of disaster research*. New York, USA: Springer, pp. 413–429.

Scacchi A. (1882) Della lava Vesuviana dell'anno 1631. *Memoire della Societa Italiana delle Scienze* IV(8): 1–34.

Schenk C.G. (2013) Navigating an inconvenient difference in antagonistic contexts: doing fieldwork in Aceh, Indonesia. *Singapore Journal of Tropical Geography* 34(3): 342–356.

Scheyvens R., Storey D. (2014) *Development fieldwork: a practical guide*. 2nd edition, London, UK: Sage.

Schrag C. (1954) Leadership among prison inmates. *American Sociological Review* 19(1): 37–42.

Schmidt J.M. (2003) Paradise lost? Social change and *fa'afafine* in Samoa. *Current Sociology* 51(3–4): 417–432.

Schmidt J.M. (2005) Migrating genders: Westernisation, migration, and Samoan *fa'afafine*. PhD thesis, The University of Auckland, Auckland, New Zealand.

Schuller M. (2012) *Killing with kindness: Haiti, international aid, and NGOs*. New Brunswick, USA: Rutgers University Press.

Schuller M. (2015) "Pa manyen fanm nan konsa": intersectionality, structural violence, and vulnerability before and after Haiti's earthquake. *Feminist Studies* 41(1): 184–210.

Schuller M., Moralles P. (2012) *Tectonic shift: Haiti since the earthquake*. Sterling, USA: Kumarian Press.

Schwartz J.A., Barry C. (2005) *A guide to preparing for and responding to prison emergencies*. Washington, DC, USA: U.S. Department of Justice.

Scott J.C. (1985) *Weapons of the weak: everyday forms of peasant resistance*. New Haven, USA: Yale University Press.

Scott J.C. (1989) Everyday forms of resistance. *Copenhagen Papers in East and Southeast Asia Studies* 4: 33–62.

Scott J.C. (1990) *Domination and the arts of resistance: hidden transcripts*. New Haven, USA: Yale University Press.

Senghor L.S. (1971) Problématique de la Négritude. *Présence Africaine* 78: 3–26.

Serje J. (2012) Data sources on hazards. In Wisner B., Gaillard J.C., Kelman I. (eds) *The Routledge handbook of hazards and disaster risk reduction*. Abingdon, UK: Routledge, pp. 179–190.

Shaw R., Dhar Chakrabarti P.G., Gupta M. (2010) *India city profile: climate and disaster resilience: consultation report*. Kyoto, Japan / New Delhi, India: Kyoto University / National Institute of Disaster Management / SEEDS India.

Shaw R., Sharma A., Takeuchi Y. (2009) *Indigenous knowledge and disaster risk reduction: from practice to policy*. New York, USA: Nova Science Publishers.

Shohat E. (1992) Notes on the "post-colonial". *Social Text* 31/32: 99–113.

Sidaway J.D. (1992) On the politics of research by 'First World' geographers in the 'Third World'. *Area* 24(4): 403–408.

Siegel J.T. (2005) Peduli Aceh. *Indonesia* 79: 165–167.

Simpson E. (2014) *The political biography of an earthquake: aftermath and amnesia in Gujarat, India.* London, UK: Hurst Publishers.

Skarbek D. (2014) *The social order of the underworld: how prison gangs govern the American penal system.* Oxford, UK: Oxford University Press.

Smith C. (2016) Inmates: our defenders in disasters. *Natural Hazards Observer* 40(8): 10–13.

Smith C. (2019) Incarcerated workers and inmate all-hazard firefighters in emergency response and disasters: a captive labor force. PhD thesis, Louisiana State University, Baton Rouge, USA.

Smith F. (2014) Gender minorities and disaster: a case study of fa'afafine in the 2012 cyclone in Samoa. BAHons dissertation, The University of Auckland, Auckland, New Zealand.

Smith N. (2008) *Uneven development: nature, capital and the production of space.* 3rd edition, Athens, USA: The University of Georgia Press.

Spence R., So E., Scawthorn C. (2011) *Human casualties in earthquakes: progress in modelling and mitigation.* Dordrecht, Netherlands: Springer.

Sphere Association (2018) *The Sphere handbook: humanitarian charter and minimum standards in humanitarian response.* 4th edition, Geneva, Switzerland: Sphere Association.

Spivak G.C. (1974) Translator's preface. In Derrida J. (Ed.) *Of grammatology.* Baltimore, USA: The Johns Hopkins University Press, pp. ix–xc.

Spivak G.C. (1985) Subaltern studies: deconstructing historiography. In Guha R. (Ed.) *Subaltern studies IV: writings on South Asian history and society.* Dehi, India: Oxford University Press, pp. 330–363.

Spivak G.C. (1987) *In other worlds: essays in cultural politics.* London, UK: Methuen Publishing.

Spivak G.C. (1988) Can the subaltern speak? In Nelson C., Grossberg L. (Eds) *Marxism and the interpretation of culture.* Urbana, USA: University of Illinois Press, pp. 271–313.

Spivak G.C. (1990a) *The post-colonial critic: interviews, strategies, dialogues.* New York, USA: Routledge.

Spivak G.C. (1990b) Theory in the margin: Coetzee's Foe reading Defoe's "Crusoe/Roxana". *English in Africa* 17(2): 1–23.

Spivak G.C. (1990c) Poststructuralism, marginality, postcoloniality and value. In Collier P., Geyer-Ryan H. (Eds) *Literary theory today.* New York, USA: Cornell University Press, pp. 219–244.

Spivak G.C. (1993a) *Outside in the teaching machine.* London, USA: Routledge.

Spivak G.C. (1993b) *The burden of English.* In Breckenridge C.A., van der Veer P. (Eds) *Orientalism and the postcolonial predicament: perspectives on South Asia.* Philadelphia, USA: University of Pennsylvania Press, pp. 134–157.

Spivak G.C. (2013) Interview with Gayatri Chakravorty Spivak. In Srivastava N., Bhattacharya B. (Eds.) *The postcolonial Gramsci.* Abingdon, UK: Routledge, pp. 233–244.

Spurlin W.J. (2006) *Imperialism within the margins: queer representation and the politics of culture in Southern Africa.* New York, USA: Palgrave Macmillan.

Stallings R.A. (2002) *Methods of disaster research.* Bloomington, USA: Xlibris.

Stallings R.A. (2007) Methodological issues. In Rodríguez H., Quarantelli E.L., Dynes R.R. (Eds) *Handbook of disaster research.* New York, USA: Springer, pp.55–83.

Stewart M.G. (2003) Cyclone damage and temporal changes to building vulnerability and economic risks for residential construction. *Journal of Wind Engineering and Industrial Aerodynamics* 91(5): 671–691.

Storey D., Hunter S. (2010) Kiribati: an environmental 'perfect storm'. *Australian Geographer* 41(2): 167–181.

Sua'ali'i T.M. (2001) Samoans and gender: some reflections on male, female and *fa'afafine* gender identities. In Macpherson C., Spoonley P., Anae M. (Eds) *Tangata o te moana nui: the evolving identities of Pacific peoples in Aotearoa/New Zealand.* Palmerston North, New Zealand: Dunmore Press, pp. 16–180.

Sumathipala A., Jafarey A., de Castro L.D., Ahmad A., Mercer D., Srinivasan S., Kumar N., Siribaddana S., Sutaryo S., Bhan A., Waidyaratne D., Beneragama S., Jayasekera C., Edirisingha S., Siriwardhana C. (2007) Ethical issues in post-disaster clinical interventions and research: a developing world perspective. *Asian Bioethics Review* 2(2): 124–142.

Sykes G.M. (1958) *The society of captives: a study of a maximum security prison.* Princeton, USA: Princeton University Press.

Tabokai N. (1993) The maneaba system. In Van Trease H. (Ed.) *Atoll politics: the Republic of Kiribati.* Christchurch, New Zealand/Suva, Fiji: Macmillan Brown Centre for Pacific Studies of the University of Canterbury/Institute of Pacific Studies of the University of the South Pacific, pp. 23–29.

Takakura H. (2019) The anthropologist as both disaster victim and disaster researcher: reflections and advocacy. In Bouterey S., Marceau E.L. (Eds) *Crisis and disaster in Japan and New Zealand.* New York, USA: Palgrave Macmillan, pp. 79–103.

Tan M.L. (1995a) From *bakla* to gay: shifting gender identities and sexual behaviors in the Philippines. In Parker R.G., Gagnon J.H. (Eds) *Conceiving sexuality: approaches to sex research in a post-modern research.* New York, USA: Routledge, pp. 85–96.

Tan M.L. (1995b) Tita Aida and emerging communities of gay men. *Journal of Gay and Lesbian Social Services* 3(3): 31–48.

Tan M.L. (2001) Survival through pluralism: emerging gay communities in the Philippines. *Journal of Homosexuality* 40(3–4): 117–142.

Tata Cissé Y., Sagot-Duvauroux J.-L. (2003) *La Charte du Mandé et autres traditions du Mali.* Paris, France: Albin Michel.

Tatel Jr. C.P. (2011) *Ethnographies of disaster: the 2009 UP anthropology field school, Tiwi, Albay.* Legazpi City, The Philippines: Aquinas University of Legazpi.

Thomas F.R. (2001) Remodeling marine tenure on the atolls: a case study from Western Kiribati, Micronesia. *Human Ecology* 29(4): 399–423.

Thomas F.R. (2002) Self-reliance in Kiribati: contrasting views of agricultural and fisheries production. *The Geographical Journal* 168(2): 163–177.

Thomas F.R., Tonganibeia K. (2007) Pacific island rural development: challenges and prospects in Kiribati. In Connell J., Waddell E. (Eds) *Environment, development and change in rural Asia-Pacific: between local and global.* London, UK: Routledge, pp. 38–55.

Thufail F.I. (2019) Between the king and the scientist: Mount Merapi eruption, early warning system, and the politics of local knowledge. In Iwamoto W., Nojima Y. (Eds) *Proceedings of the Asia-Pacific Regional Workshop on Intangible Cultural Heritage and Natural Disasters – 7-9 December 2018, Sendai, Japan.* Osaka, Japan: International Research Centre for Intangible Cultural Heritage in the Asia-Pacific Region, pp. 60–64.

Tierney K., Bevc C., Kuligowski E. (2006) Metaphors matter: disaster myths, media frames, and their consequences in Hurricane Katrina. *The Annals of the American Academy of Political and Social Science* 604(1): 57–81.

Timmerman P. (1981) *Vulnerability, resilience and the collapse of society: a review of models and possible climatic applications.* Environmental Monograph No. 1, Institute for Environmental Studies, University of Toronto, Toronto.

Tindowen D.J.C., Bagalayos H.L.N. (2018.) All for one, one for all: the role of Filipino pro-social behaviours in building a disaster-resilient community. *International Journal of Sustainable Society* 10(3): 243–256.

Titz A., Cannon T., Krüger F. (2018) Uncovering 'community': challenging and elusive concept in development and disaster related work. *Societies* 8(71): doi: 10.3390/soc8030071

Tönnies F. (1887) *Gemeinschaft and Gesellschaft.* Leipzig, Germany: Fues's Verlag.

Torry W.I. (1979a) Intelligence, resilience and change in complex social systems: famine administration in India. *Mass Emergencies* 2: 71–85.

Torry W.I. (1979b) Anthropological studies in hazardous environments: past trends and new horizons. *Current Anthropology* 20(3): 517–540.

Travers M. (2019) The idea of Southern criminology. *International Journal of Comparative and Applied Criminal Justice* 43(1): 1–12.

Tsunozaki E. (2012) Community-based disaster risk reduction through 'town-watching' in Japan. In Wisner B., Gaillard J.C., Kelman I. (Eds) *Handbook of hazards and disaster risk reduction.* Abingdon, UK: Routledge, pp. 719–720.

Tuhiwai-Smith L. (2012) *Decolonizing methodology: research and indigenous people.* 2nd edition, London, UK: Zed Books.

Turnbull M., Sterrett C.L., Hilleboe A. (2013) *Toward resilience: a guide to disaster risk reduction and climate change adaptation.* Rugby, UK: Practical Action Publishing.

Turner II B.L., Kasperson R.E., Matson P.A., McCarthy J.J., Corell R.W., Christensen L., Eckley N., Kasperson J.X., Luers A., Martello M.L., Polsky C., Pulsipher A., Schiller A. (2003) A framework for vulnerability analysis in sustainability science. *Proceedings of the National Academy of Sciences of the United States of America* 110(14): 8074–8079.

Turner S. (2010) The silenced assistant. Reflections of invisible interpreters and research assistants. *Asia Pacific Viewpoint* 51(2): 206–219.

Twigg J. (2009) *Characteristics of a disaster-resilient community: a guidance note.* Version 2, Aon Benfield UCL Hazard Research Centre, London, UK: University College London.

United Nations Development Programme Drylands Development Centre (2014) *Community Based Resilience Analysis (CoBRA) conceptual framework and methodology.* New York, USA: United Nations Development Programme Drylands Development Centre.

United Nations Framework Convention on Climate Change (2015) *Paris agreement.* Bonn, Germany: United Nations Framework Convention on Climate Change.

United Nations International Strategy for Disaster Reduction (2002) *Living with risk: a global review of disaster reduction initiatives.* Geneva, Switzerland: United Nations International Strategy for Disaster Reduction.

United Nations International Strategy for Disaster Reduction (2009) *2009 global assessment report on disaster risk reduction: risk and poverty in a changing climate.* Geneva, Switzerland: United Nations International Strategy for Disaster Reduction.

United Nations International Strategy for Disaster Reduction (2015) *Sendai framework for disaster risk reduction 2015–2030.* Geneva, Switzerland: United Nations Office for Disaster Risk Reduction.

United Nations International Strategy for Disaster Reduction, United Nations Development Programme, and International Union for Conservation of Nature (2011) *Making disaster risk reduction gender-sensitive: policy and practical guidelines*. Geneva, Switzerland: United Nations International Strategy for Disaster Reduction / United Nations Development Programme / International Union for Conservation of Nature.

United Nations Office for Disaster Risk Reduction (2019a) *2019 global assessment report on disaster risk reduction*. Geneva, Switzerland: United Nations Office for Disaster Risk Reduction.

United Nations Office for Disaster Risk Reduction (2019b) *Words into action: developing national disaster risk reduction strategies*. Geneva, Switzerland: United Nations Office for Disaster Risk Reduction.

United Nations Office for Disaster Risk Reduction (2020) *Words into action: engaging children and youth in disaster risk reduction and resilience building*. Geneva, Switzerland: United Nations Office for Disaster Risk Reduction.

Vaioleti T.M. (2006) Talanoa research methodology: a developing position on Pacific research. *Waikato Journal of Education* 12: 21–34.

Valdés H.M. (1995) Expanding women's participation in disaster prevention and mitigation: some approaches from Latin America and the Caribbean. *Stop Disasters* 24: 10–11.

Vaughan M. (1987) *The story of an African famine: gender and famine in twentieth-century Malawi*. Cambridge, UK: Cambridge University Press.

Verges F. (2019) *Un féminisme décolonial*. Paris, France: La Fabrique.

Vickery J. (2018) Using an intersectional approach to advance understanding of homeless persons' vulnerability to disaster. *Environmental Sociology* 4(1): 136–147.

Vietnam Red Cross Society (2000) *Disaster preparedness manual*. Hanoi, Vietnam: Vietnam Red Cross Society.

Waddell E. (1977) The hazards of scientism: a review article. *Human Ecology* 5(1): 69–76.

Wanandi J. (2002) Islam in Indonesia: its history, development and future challenges. *Asia-Pacific Review* 9(2): 104–112.

Waru R. (2012) *Secrets and treasures: our stories through the objects at Archives New Zealand*. Auckland, New Zealand: Random House.

Watson J.B. (1913) Psychology as the behaviourist views it. *Psychological Review* 20(2): 158–177.

Watters R. (2008) *Journeys towards progress: essays of a geographer on development and change in Oceania*. Wellington, New Zealand: Victoria University Press.

Watts M.J., Bohle H.G. (1993) The space of vulnerability: the causal structure of hunger and famine. *Progress in Human Geography* 17(1): 43–67.

Webb G.R. (2007) The popular culture of disaster: exploring a new dimension of disaster research. In Rodríguez H., Quarantelli E.L., Dynes R.R. (Eds) *Handbook of disaster research*. New York, USA: Springer, pp. 430–440.

Webber M.M. (1964) Urban place and the non-place urban realm. In Webber M.M. (Ed.) *Explorations into urban structure*. Philadelphia, USA: University of Pennsylvania Press, pp. 79–153.

Webber S. (2013) Performative vulnerability: climate change adaptation policies and financing in Kiribati. *Environment and Planning A* 45(11): 2717–2733.

Weichselgartner J., Kelman I. (2015) Geographies of resilience: challenges and opportunities of a descriptive concept. *Progress in Human Geography* 39(3): 249–267.

Wenger D.E. (1978) Community response to disaster: functional and structural alternations. In Quarantelli E.L. (Ed.) *Disasters: theory and research*. London, UK: Sage Publications, pp. 17–47.

Wenger D.E., Weller J.M. (1973) *Disaster subcultures: the cultural residues of community disasters*. Preliminary Paper No. 9, Disaster Research Center, Newark, USA: University of Delaware.

Werner E.E., Bierman J.M., French F.E. (1971) *The children of Kauai: a longitudinal study from the prenatal period to age ten*. Honolulu, USA: University of Hawaii Press.

Westphal C. (1870) Die conträre sexualempfindung, symptom eines neuropathischen (psychopathischen) zustandes. *Archiv für Psychiatrie und Nervenkrankheiten* 2: 73–108.

White G.F. (1945) *Human adjustment to floods: a geographical approach to the flood problem in the United-States*. Research Paper No 29, Chicago, USA: Department of Geography-University of Chicago.

White G.F. (1974) *Natural hazards: local, national, global*. New York, USA: Oxford University Press.

White G.F., Haas J.E. (1975) *Assessment of research on natural hazards*. Cambridge, USA: The MIT Press.

White S.C. (1996) Depoliticising development: the uses and abuses of participation. *Development in Practice* 6(1): 6–15.

Williams G. (2004) Evaluating participatory development: tyranny, power and (re)politicisation. *Third World Quarterly* 25(3): 557–578.

Williams R. (1960) *Culture & society 1780–1950*. New York, USA: Anchor Books.

Winderl T. (2014) *Disaster resilience measurements: stocktaking of ongoing efforts in developing systems for measuring resilience*. New York, USA: United Nations Development Programme.

Windle G., Bennet K.M., Noyes J. (2011) A methodological review of resilience measurement scales. *Health and Quality of Life Outcomes* 9(8): doi: 10.1186/1477-7525-9-8.

Wiredu K. (1980) *Philosophy and an African culture*. Cambridge, UK: Cambridge University Press.

Wirtz A., Kron W., Löw P., Steuer M. (2014) The need for data: natural disasters and the challenges of database management. *Natural Hazards* 70(1): 135–157.

Wisner B. (1993) Disaster vulnerability: scale, power, and daily life. *Geojournal* 30(2): 127–140.

Wisner B. (2001) Risk and the neoliberal state: why post-Mitch lessons didn't reduce El Salvador's earthquake losses. *Disasters* 25(3): 251–268.

Wisner B. (2006) *Let our children teach us! A review of the role of education and knowledge in disaster risk reduction*. Geneva, Switzerland: United Nations International Strategy for Disaster Reduction.

Wisner B. (2016) Vulnerability as concept, model, metric, and tool. In Cutter S.L. (Ed.) *Oxford research encyclopedias: natural hazard science*. New York, USA: Oxford University Press. Available from http://naturalhazardscience.oxfordre.com/view/10.1093/acrefore/9780199389407.001.0001/acrefore-9780199389407-e-25 (accessed 6 June 2018).

Wisner B., Gaillard J.C. (2009) An introduction to neglected disasters. *Jàmbá: Journal of Disaster Risk Studies* 2(3): 151–158.

Wisner B., Westgate K., O'Keefe P. (1976) Poverty and disaster. *New Society* 37(727): 546–548.

Wisner B., O'Keefe P., Westgate K. (1977) Global systems and local disasters: the untapped power of peoples' science. *Disasters* 1(1): 47–57.

Wisner B., Blaikie P., Cannon T, Davis I. (2004.) *At risk: natural hazards, people's vulnerability, and disasters*. 2nd edition, London, UK: Routledge.

Wisner B., Gaillard J.C., Kelman I. (2012) *Handbook of hazards and disaster risk reduction*. Abingdon, UK: Routledge.

Wittig M. (1980) On ne naît pas femme. *Questions Féministes* 8: 75–84.

Wittig M. (1992) *The straight mind and other essays.* Boston, USA: Beacon Press.

Wood G. (1985) The politics of development policy labelling. *Development and Change* 16(3): 347–373.

World Bank (The) (2010) *Reducing the risk of disasters and climate variability in the Pacific Islands: Republic of Kiribati country assessment.* Washington, DC, USA: The World Bank.

World Bank (The), United Nations (The) (2010) *Natural hazards, unnatural disasters: the economics of effective prevention.* Washington, DC, USA: The World Bank.

Yiftachel O. (1998) Planning and social control: exploring the dark side. *Journal of Planning Literature* 12(4): 395–406.

Young R.J.C. (1995a) Foucault on race and colonialism. *New Formations* 25: 57–65.

Young R.J.C. (1995b) *Colonial desire: hybridity in theory, culture and race.* London, UK: Routledge.

Young R.J.C. (2004) *White mythologies: writing history and the West.* 2nd edition, London, UK: Routledge.

Zuñiga M.L. (2007) *Priorities of the people: hardship in Kiribati.* Manila, The Philippines: Asian Development Bank.

Index

Printed in the United States
by Baker & Taylor Publisher Services

Printed in the United States
by Baker & Taylor Publisher Services